Linking EU Climate and Energy Policies

NEW HORIZONS IN ENVIRONMENTAL POLITICS

Series Editor: Arthur Mol, *Chair and Professor in Environmental Policy, Director, Wageningen School of Social Sciences, Wageningen University, the Netherlands, Professor in Environmental Policy, Renmin University, Beijing*

The New Horizons in Environmental Politics series provides a platform for in-depth critical assessments of how we understand the many changes in the politics of nature, the environment and natural resources that have occurred over the last 50 years. Books in the series question how the environment is (re)defined, debated and protected; explore differences between countries and regions in environmental politics; analyse how actors do and do not collaborate around environment and natural resource conflicts; describe who wins and who loses and in what ways; and detail how to better study, analyse and theorize such developments and outcomes.

The series is designed to promote innovative cross-disciplinary analysis of the contemporary issues and debates influencing the various dimensions of environmental politics. Covering a diverse range of topics, the series will examine the political, economic and ethical aspects of environmental policy, governance and regulation. It brings together cutting edge research on environmental politics worldwide in order to shed light on, and explain current trends and developments.

With oversight from the Series Editor, Professor Arthur Mol – a noted specialist in the field of environmental politics at Wageningen University, the Netherlands – the New Horizons in Environmental Politics series comprises carefully commissioned projects from experts in the field including both academics and professionals. The audience for the series is global, and books in the series are essential reading for students, academics and professionals – in short, anyone with an interest in understanding the vital issues affecting environmental politics in the twenty-first century.

Titles in the series include:

Russia and the Politics of International Environmental Regimes
Environmental Encounters or Foreign Policy?
Anna Korppoo, Nina Tynkkynen and Geir Hønneland

American Environmental Policy
The Failures of Compliance, Abatement and Mitigation
Daniel Press

Environmental Governance through Partnerships
A Discourse Theoretical Study
Ayşem Mert

The Domestic Politics of Global Climate Change
Key Actors in International Climate Cooperation
Edited by Guri Bang, Steinar Andresen and Arild Underdal

Linking EU Climate and Energy Policies
Decision-making, Implementation and Reform
Jon Birger Skjærseth, Per Ove Eikeland, Lars H. Gulbrandsen and Torbjørg Jevnaker

Linking EU Climate and Energy Policies

Decision-making, Implementation and Reform

Jon Birger Skjærseth

Research Professor, Fridtjof Nansen Institute, Lysaker, Norway

Per Ove Eikeland

Senior Research Fellow, Fridtjof Nansen Institute, Lysaker, Norway

Lars H. Gulbrandsen

Research Professor, Fridtjof Nansen Institute, Lysaker, Norway

Torbjørg Jevnaker

Research Fellow, Fridtjof Nansen Institute, Lysaker, Norway

NEW HORIZONS IN ENVIRONMENTAL POLITICS

Edward Elgar PUBLISHING

Cheltenham, UK • Northampton, MA, USA

Published by
Edward Elgar Publishing Limited
The Lypiatts
15 Lansdown Road
Cheltenham
Glos GL50 2JA
UK

Edward Elgar Publishing, Inc.
William Pratt House
9 Dewey Court
Northampton
Massachusetts 01060
USA

A catalogue record for this book
is available from the British Library

Library of Congress Control Number: 2015959657

This book is available electronically in the **Elgar**online
Social and Political Science subject collection
DOI 10.4337/9781785361289

MIX
Paper from
responsible sources
FSC® C013056

ISBN 978 1 78536 127 2 (cased)
ISBN 978 1 78536 128 9 (eBook)

Typeset by Columns Design XML Ltd, Reading
Printed and bound in Great Britain by TJ International Ltd, Padstow

Contents

List of tables vi
Preface and acknowledgements vii
List of abbreviations ix

1 Introduction 1
Per Ove Eikeland and Jon Birger Skjærseth
2 Analytical framework 11
Jon Birger Skjærseth and Per Ove Eikeland
3 Evolution of EU climate and energy policies 31
Per Ove Eikeland and Jon Birger Skjærseth
4 Initiating the package for 2020 50
Per Ove Eikeland, Jon Birger Skjærseth and Lars H. Gulbrandsen
5 Deciding the package for 2020 71
Jon Birger Skjærseth, Per Ove Eikeland and Lars H. Gulbrandsen
6 Implementation in Germany 91
Per Ove Eikeland
7 Implementation in Poland 120
Jon Birger Skjærseth
8 Implementation in the Netherlands 147
Lars H. Gulbrandsen and Jon Birger Skjærseth
9 Implementation in Norway 175
Torbjørg Jevnaker
10 Comparative analysis and consequences for EU 2030 204
Jon Birger Skjærseth, Per Ove Eikeland, Lars H. Gulbrandsen and Torbjørg Jevnaker
11 Conclusions and the road ahead 241
Jon Birger Skjærseth and Per Ove Eikeland

Interviews 252
Index 257

Tables

2.1	Summary of relationship between implementation experiences and reform	24
2.2	Theories and policy phases	26
10.1	Implementation of the package by 2014 with a view to 2020	208
10.2	Fit and adaptation pressure: correspondence between expectations and observations	213
10.3	Domestic politics explanations of implementation problems	223
10.4	Implementation experiences and long-term positions	226
10.5	Share of renewables in total energy consumption and net energy-import dependency, 2005/2012	227

Preface and acknowledgements

2008: as political scientists, we were puzzled – how could the 27 diverse EU member-states manage to agree on climate and energy policies for 2020 that departed significantly from the preferences of the least ambitious actors? Traditional theories of EU integration and policymaking did not suffice to explain this development, so we opted for a different approach: integrating negotiation theories on issue- and policy-linkage to get a better grip on how different pivotal interests and values might be combined to promote policy agreement when unanimity is required. Additionally, we were curious about how the EU's climate and energy policies would be implemented in the member-states, and in Norway as a European Economic Agreement country. In 2010, we were fortunate in receiving a project grant from the Research Council of Norway that enabled closer study of these issues.

As our project progressed, the EU commenced negotiations on new climate and energy policies for 2030. This gave us the opportunity to study the whole policy-cycle – from the initiation of 2020 policies, to reform for 2030 – in a coherent fashion. One 'take-home' message from this book is that linking climate and energy policies in one phase affects outcomes in the next phase, as is evident from how implementation experiences have affected national positions on new long-term climate policies. Another message is that combining traditional EU theories with negotiation theory can explain why and how policies develop within and across the various phases of initiation, decision-making, implementation and reform. Luckily, we were able to explore also the reform phase, thanks to additional funding from the Research Council of Norway's grant to CICEP – a centre for collaborative research on strategic challenges in international climate and energy policy, involving the Centre for International Climate and Environmental Research – Oslo (CICERO), the Department of Political Science at the University of Oslo, and the Fridtjof Nansen Institute (FNI).

Lars H. Gulbrandsen and Torbjørg Jevnaker had responsibility for various empirical parts of the project, and this is reflected in differing author responsibilities for the individual chapters. At the same time, this book is very much the result of joint efforts involving the whole project group, reflected in joint authorship. Specifically, this volume builds on various articles and reports that have been presented and disseminated to national and international audiences along the way.

Many scholars have provided extremely valuable inputs and constructive criticism. Anne Raaum Christensen was part of the initial project team; she has contributed on EU climate policy in the transport sector. Steinar Andresen and Olav Schram Stokke have provided input to the analytical framework. Arild Underdal, Guri Bang, Inga Ydersbond and Miranda Schreurs have commented on related work that has fed into this book. Our special thanks go to Andy Jordan, Harro van Asselt and Sebastian Oberthür, for reviewing portions of the manuscript. We are as always grateful to Susan Høivik, who has improved the English text considerably. Finally, we wish to thank representatives from EU institutions and national governments as well as interest organizations, for their openness and for taking the time to talk with us.

Jon Birger Skjærseth and Per Ove Eikeland
October, 2015
Lysaker, Norway

Abbreviations

ACEA	European Automobile Manufacturers Association
AEA	Annual Emission Allocation
ALTENER	European programme on renewable energy sources
BDI	Federation of German Industries
BEE	German Renewable Energy Industry Association
BMU	Bundesministerium für Umwelt (Ministry for the Environment) (Germany)
BMWi	Bundesministerium für Wirtschaft und Energie (Ministry of Economics and Energy) (Germany)
CCS	Carbon capture and storage
CDM	Clean Development Mechanism
CDU	Christian Democratic Union (Germany)
CEECs	Central and Eastern European countries
CHP	Combined heat and power
Commission	European Commission
CoP	Conference of the Parties (UNFCCC Treaty)
CSU	Christian Social Union (Germany)
DG TREN	Directorate-General for Energy and Transport (EU)
ECCP	European Climate Change Programme
EEA	European Economic Area
EED	Energy Efficiency Directive
EFTA	European Free Trade Association
EIA	Energy investment tax deduction (the Netherlands)
Energy Agreement	Energy Agreement for Sustainable Growth (the Netherlands)
EREC	European Renewable Energy Council
ESA	EFTA Surveillance Authority
ESD	Effort-Sharing Decision
ETS	Emissions Trading System
EU ETS	EU Emissions Trading System
EU-15	EU members 1995–2004
EV	Electric vehicle

FDP	Free Democratic Party (Germany)
FNI	Fridtjof Nansen Institute
FQD	Fuel Quality Directive
GHG	Greenhouse gas
GW	Gigawatts
HLG	High-Level Group
HSE	Health, safety, environment
IEA	International Energy Agency
IECP	Integrated Energy and Climate Programme (Germany)
ILUC	Indirect land-use change
IPCC	Intergovernmental Panel on Climate Change
ITRE Committee	Industry, Research and Energy Committee (EU Parliament)
JI	Joint Implementation
KfW	Kreditanstalt für Wiederaufbau (Germany)
LI	Liberal Intergovernmentalism
LNG	Liquefied natural gas
LTA3	Long-Term Agreement on Energy Efficiency for non-ETS enterprises (the Netherlands)
LULUCF	Land use, and land-use change and forestry
MLG	Multi-level Governance
MoE	Ministry of the Environment (Norway)
MPE	Ministry of Petroleum and Energy (Norway)
MSR	Market Stability Reserve
NAPs	National Allocation Plans
NCPIP	National Climate Policy Implementation Plan (the Netherlands)
NER	New Entrants' Reserve (EU ETS)
NGO	Non-governmental organization
NIM	National implementation measure
NREAP	National Renewable Energy Action Plan
PBL	Dutch environmental assessment agency (formerly MNP)
PiS	Law and Justice Party (Poland)
PJ	Petajoule
PSL	Polish People's Party
PV	Photovoltaic
RED	Renewable Energy Directive
SAVE	Specific Actions for Vigorous Energy Efficiency (EU)

SDE, SDE-plus	Stimulering Duurzame Energieproductie/ Encouraging Sustainable Energy Production, an operating grant/subsidy (the Netherlands)
SER	Social and Economic Council (the Netherlands)
SPD	Social Democratic Party (Germany)
Storting	Norwegian Parliament
TEHG	Greenhouse Gas Emissions Trading Act (Germany)
UNCED	UN Conference on Environment and Development
UNFCCC	UN Framework Convention on Climate Change
UNICE	Union of Industrial and Employers' Confederations of Europe (now Business Europe)
VDMA	German Engineering Association
VNO-NCW	Confederation of Netherlands Industry and Employers
VROM	Ministry of Housing, Spatial Planning and the Environment (the Netherlands)
V-4	Visegrad Group

1. Introduction

Per Ove Eikeland and Jon Birger Skjærseth

The EU has aimed to play a leadership-by-example role in international climate negotiations, adding increasingly ambitious targets and policy instruments. Its climate ambitions reached a new peak in October 2014, when the 28 heads of states and governments agreed to reduce greenhouse gas (GHG) emissions within the EU by at least 40 per cent between 1990 and 2030. This had been preceded by the March 2007 agreement among EU leaders to reduce GHGs by at least 20 per cent between 1990 and 2020, followed one year later by the unanimous adoption of a package of binding climate and energy policies. The chances for realizing the EU's long-term decarbonization goal by the year 2050 thereby gained traction.[1]

This leadership ambition in climate policy gives rise to several intriguing questions from a political science perspective. How did the various member-states manage to agree on increasingly ambitious common targets and policies? Have the policies been implemented domestically – and have they served to trigger a transformation towards decarbonization? What are the consequences of these implementation experiences for the adoption of new EU policies to 2030 and beyond? The learning effect of the EU experience may prove considerable, as leaders around the world must grapple with related political feasibility challenges in responding to the threat of climate change (see Intergovernmental Panel on Climate Change (IPCC) 2014). Can climate policies be strengthened without creating losers who may try to block subsequent policy development? Can policies be designed in ways that promote domestic implementation and lower-carbon transformations? And how can policymakers reform and step up existing policies to meet the long-term challenge of climate change? Even though the EU is a 'unique' political system confronting distinctive energy challenges, its experience can be valuable for other factors, highlighting opportunities and pitfalls in responding to the problem of climate change.[2]

This book explains policymaking, implementation and reform by combining negotiation theory on issue-linkages with theories on EU policymaking and implementation. Using a novel analytical approach we follow the policy-cycle from the initiation, decision-making and implementation of targets and policies for 2020, to reform for 2030. The book offers insights into how climate ambitions can be raised under the most demanding decision-making rule – when unanimity is required. In the main, EU climate and energy targets and policies have been adopted by unanimity, which has alternately been legally required, politically determined or *de facto* under the 'shadow of voting'. Negotiation theory has shown that agreement can be reached through *issue-linking*, by designing policy packages that combine different actor interests and values, and promote synergies (Sebenius 1983). However, policy packages may also lead to policy 'trouble'. Bringing 'wrong' issues into policy packages may make negotiations too complex, splitting rather than unifying actor interests – and eventually failing.

We combine negotiation theory on issue-linkages with theories of EU integration, policymaking and implementation. Negotiations are necessary to the functioning of the EU. The once-distinct literatures on EU politics and negotiations theory are increasingly interlinked in the study of EU integration (Dür et al. 2010). Still, as far as we are aware, applying this combination of theories to climate and energy policies, encompassing the full policy-cycle, is unprecedented.[3]

And finally, we show the value added of tracking politics and policy development over the full range of policy phases – from initiation, through decision-making, to implementation and policy reform. This dynamic approach reveals how policies and issue-linkages within one phase affect outcomes in the next phase. Reform of existing EU policies for 2020, as illustrated by the EU 2030 climate- and energy-policy framework, is affected by how policies were initially made and by experiences gained from domestic implementation. Crafting 'successful' policy packages requires skilful entrepreneurship over time – not merely as a one-shot effort.

Our starting point in seeking to explain the initiation and adoption of the EU's 2020 targets and policies is the following paradox: EU climate policy comes close to the positions of the 'most ambitious' actors even when the decisions have been adopted by unanimity. This unanimity included the ten Central and Eastern European countries (CEECs) that started to enter the EU from 2004 – some of them, such as Poland, highly sceptical towards a more ambitious EU climate policy. According to the null hypothesis indicated by negotiation theory, EU climate policy would reflect the position of the 'least ambitious' actor when unanimity is required (Underdal 1980). As this expectation diverged fundamentally from realities, we began searching for mechanisms that could explain how all member-states could agree to policies that were significantly at odds with the interests and preferences of the 'least ambitious' actors. In theories on issue-linkage we found one compelling answer: agreement can be reached by combining different issues in a policy package. We observed that a key mechanism in EU policy-linking was to combine issues that were differently valued among pivotal actors who could block policy progress, issues such as differing concerns for climate change and energy security. Another mechanism involved compensating the 'losers' by adding new issues as side-payments, such as revenues from auctioning of emissions-trading allowances. A third mechanism was to create synergies by which policies for combating climate change could reduce air pollution and create new 'green' jobs in low-carbon energy industries, thereby raising the willingness to agree on climate policies.

Negotiation theory proves necessary for understanding EU policy progress under conditions of unanimity, but theories of EU integration and policymaking are needed to explain why and how policies developed in the first place. Here, we apply two approaches: Liberal Intergovernmentalism (LI) and Multi-level Governance (MLG). From the LI perspective, we would expect the initiation and adoption of climate and energy policies to reflect the interests, preferences and actions of the member-states and their intergovernmental bargaining (Moravcsik 1998, 1999). Entrepreneurial member-states can take the lead in crafting new and ambitious package solutions because they believe that issue-linkage will improve the chances that less ambitious member-states will agree to more ambitious EU policies.

LI has been challenged by other approaches that hold that European integration includes other actors, institutions and channels for influence that challenge member-states. MLG outlines various mechanisms that restrain member-state governments from having full control (Hooghe and Marks 2001). Two such mechanisms are particularly relevant here. First, supranational institutions have an independent influence on policymaking as actors and arenas, an influence beyond the scope of their role as agents serving national governments. Secondly, non-state actors may directly influence both EU and national policymaking (Skjærseth and Wettestad 2008). These mechanisms can work in combination. The European Commission (henceforth Commission) has the power to initiate new legislation and can serve as a political entrepreneur by taking independent initiatives within 'networks' of non-state actors seeking to exert influence on EU legislation (Richardson 1996). In essence, this MLG approach may explain the initiation and adoption of EU climate and energy policies as a result of complex bargaining at multiple governance levels that include EU institutions, non-state actors and member-state governments.

Implementation theory, derived from the MLG literature, provides our starting point for explaining domestic implementation of the EU package of binding climate and energy policies for 2020. We assess and explain implementation in selected national contexts (see below). We employ two main approaches focusing on the relationship between the EU and member-state levels to explain variation in implementation (Treib 2008; Di Lucia and Kronsell 2010). The first centres on the degree of 'fit' between EU requirements and national status quo. A good match will increase the feasibility of implementation. The chances of successful implementation would be particularly good if policies adopted at the EU level were linked or combined so as to promote benefits for all member-states. When unanimity is required at the EU level and policies are based on policy-package negotiations, however, there is also a pitfall as decision-makers can trade concessions between the various package components which, at the domestic level, might distort the 'fit' and protract later implementation of specific policies.

The second approach focuses on the relationship between the EU and domestic politics. Domestic politics explains responses to EU policies by the state (or its government) and society, and the relationship between the state and society (Underdal and Hanf 2000; Treib 2008; Di Lucia and Kronsell 2010). New governments may bring in new priorities after EU policies have been negotiated and adopted. Societal actors may be strengthened or weakened by new EU policies, and policy style may affect these actors' access to and influence on decision-making. Finally, administrative organization may promote or inhibit EU implementation.

Domestic implementation of short-term policies is likely to affect national positions on new long-term policies. How have implementation experiences affected the development of new longer-term targets and policies? Reform towards EU 2030 climate and energy targets and policies will probably reflect experiences and learning from implementing current policies for 2020. Positive experiences from implementation in the form of new low-carbon opportunities are likely to spur support for more ambitious targets based essentially on strengthening the linking of the same set of policies and issues. Conversely, negative experiences are likely to spur opposition to more of the same policies; and mixed experiences are likely to lead to preferences for re-packing, strengthening positively experienced policy elements and weakening negative elements.

Common to both LI and MLG approaches is the emphasis on EU internal factors and processes. However, international events and regimes external to the EU can also influence the development, implementation and reform of EU policies. Because the EU is predominantly an import region for energy, it is vulnerable to events affecting supply and energy prices in the international energy markets. International events may affect the political priority accorded to climate and energy issues within the EU, and hence how policies are initiated and adopted. Further, the EU, with its member-states, seeks to harmonize international commitments across the globe, in order to create a level playing field as regards competition conditions for its industries operating in international markets. Therefore, also international regimes can affect the development of EU policies (Costa and Jørgensen 2012). For one thing, harmonization of international climate commitments may reduce opposition from European industries concerning more ambitious EU climate policies. Also, international flexibility mechanisms can reduce compliance costs within the EU. Accordingly, changes in the international climate regime can affect actor positions and the adoption of new EU policies. Our theory framework is further outlined in Chapter 2.

RESEARCH DESIGN

To explain how the EU has been able to adopt, unanimously, increasingly ambitious climate policies by linking energy issues and policies, we distinguish between functional and political linkages. Functional linkages refer to trade-offs or synergies between climate and energy policies. Climate and energy issues are functionally linked: energy policies will affect GHG emissions, and climate policies will affect energy production and use. Political linking is defined according to whether climate and energy policies are initiated, proposed and adopted at the same time by the same set of policymakers. Climate and energy policies are considered de-linked or decoupled to the extent they are functionally connected but have been initiated, proposed and adopted separately at different points in time.

To explain how EU policies have been implemented and reformed, we have selected four countries for in-depth study: Germany, the Netherlands, Norway and Poland. The first criterion for selecting these cases is that they provide significant variation in energy mix and import dependency well before EU climate *and* energy targets and policies were linked at the EU level (Eurostat 2007). This means they are confronted with a range of different challenges and opportunities as regards decarbonizing their economies.

In 2005, the Netherlands imported nearly 40 per cent of its energy needs, with energy consumption based almost exclusively on fossil fuels, mainly natural gas. Norway was a net energy exporter based almost exclusively on renewable hydropower for electricity production, which accounted for about 40 per cent of gross energy consumption. Poland imported 18 per cent of its energy needs; and 90 per cent of its electricity production was based on coal, representing nearly 60 per cent of its energy consumption. Germany was the most import-dependent (61 per cent) of the four, with a mix of coal (24 per cent), oil and petroleum (36 per cent), natural gas (23 per cent), nuclear (12 per cent) and renewables (5 per cent) in its gross energy consumption.

Secondly, these four countries represent the main differences in positions in the reform negotiations on EU climate and energy policies for 2030 (see Chapter 10). Poland has served as the informal leader of a larger group of 'least enthusiastic' CEECs in the EU. Germany, the Netherlands and Norway have all been part of the 'Green Growth Group' arguing for differing but more ambitious EU goals and policies. Our third criterion has been a pragmatic one. The authors of the various country chapters are all scholars who are familiar with the climate and energy policies of these countries.

Norway provides an opportunity to explore implementation of EU policies in a country that is not an EU member. However, it is part of the single market through the European Economic Area (EEA) agreement and is thus obliged to implement EU climate and energy legislation more or less like EU member-states (see Chapter 9). On the other hand, Norway has no seat at the decision-making table in Brussels, which makes it a special case as regards the relationship between policymaking and implementation.

We have structured the four implementation studies according to a common template. They start with an account of climate and energy policies before the EU policy package negotiations in 2007 and 2008, followed by an assessment of each country's role and positions in the negotiations as compared against the EU policy outcomes. We then explore the implementation of various EU climate and energy policies with a view to 2020 and beyond. The case studies assess implementation of the 'core' package, which consists of the Directive reforming the EU Emissions Trading System (EU ETS), the Effort-Sharing Decision (ESD), the Renewable Energy Directive (RED) and the Carbon Capture and Storage (CCS) Directive. Implementation of climate policies for the transport sector is analysed only for a sub-set of countries that are particularly affected. Implementation of the car emissions regulation is analysed for Germany, with its large car manufacturing sector, and the

Fuel Quality Directive (FQD) for Norway, with its sizeable petroleum industry. Finally, the case studies assess experiences with implementation of these policies, and the related consequences for national positions on EU 2030 policies.

The disadvantage of such a structured comparative approach is that the case studies are less context-sensitive than they would have been if studied in isolation. We have tried to compensate for this by encouraging each author to identify explanatory factors that are not part of the framework, and by applying the framework with flexibility to adapt to unique case features, such as the German federal political system or Poland's post-Soviet 'culture'.

Caution is warranted in assessing implementation with a view to 2020 and beyond. Major criticism has been raised against studies of public policy implementation where large programmes have been evaluated after only a few years of operation (Skjærseth 2000). Changing behaviour and transforming energy systems take many years, and implementation may be a matter of incremental change. Still, the EU member-states (and Norway) were required to implement the package of policies by a certain deadline – so at least the policies and measures should be in place today, although the impact on emissions and energy patterns may emerge gradually, over time.

Our analysis is based on combining historical records and qualitative case studies at the EU and national levels. We extract empirically assessable propositions from the various explanatory perspectives, and then examine them by a combination of pattern-matching and explanation-building. Pattern-matching is our preferred methodological technique for assessing well-established theories, such as 'fit' and adaptation pressure. Explanation-building based on 'narratives' and process-tracing is used for capturing more complex causal relationships. Data collection has been based on multiple sources, including energy statistics, position papers, secondary literature and interviews. Semi-structured interviews have been conducted with carefully selected representatives from EU institutions, non-state actors at the EU level, and representatives of various state and private organizations and interests in the four case-study countries. Interviewees are listed at the end.

OUTLINE OF THIS BOOK

The next chapter presents our analytical framework that we use to assess and explain the initiation, adoption, implementation and reform of EU climate and energy targets and the 2008 policy package. This chapter guides the empirical analysis. Chapters 3 to 5 examine the initiation and adoption of EU policies. Chapter 3 provides some historical background to EU climate and energy policies and a baseline from which the linking of climate and energy targets and policies is studied. Chapter 4 focuses on the initiation of the climate and energy policies until the formal package proposal was presented by the Commission in January 2008. Chapter 5 directs the spotlight on the negotiations on the formal package proposal, and the outcomes and subsequent policy development at EU level until 2014. Chapters 6 to 9 analyse the implementation of EU climate and energy policies, starting with Germany in Chapter 6. The EU's renewable electricity production giant is struggling to implement several parts of the climate and energy package. Chapter 7 focuses on Poland, showing how and why the EU's sixth-largest member-state in terms of population has systematically opposed and then absorbed EU climate and energy policies. Chapter 8 turns to the Netherlands, a traditional environmental frontrunner that has shown surprisingly little interest in renewables. Chapter 9 examines the EU 'outsider', Norway, noting the challenges of combining petroleum production and an initially high level of renewables with an ambitious climate policy.

In Chapter 10, we compare implementation across the four countries and different policies. First we compare implementation of EU policies with a view to 2020 and beyond; then turn to explain variation in implementation. This is followed by analyses of the consequences of implementation experiences for national positions on the new EU 2030 climate and energy framework. Lastly, we feed the various experiences into an analysis of the adoption of the EU 2030 climate- and energy-policy framework in October 2014. Our empirical findings, with analytical implications and views on the road ahead, are presented in the concluding Chapter 11.

NOTES

1. In October 2009, EU leaders agreed to support the EU goal of reducing GHG emissions by between 80 and 95 per cent by the year 2050 compared to 1990 levels (European Council 2009).
2. The EU is the world's largest energy importer, its energy-import dependency has been rising since the mid-1990s. In 2012, 53 per cent of EU energy consumption was linked to imports. Energy imports to the EU cost about one billion euros per day; EU energy dependency is 88 per cent of crude oil, 66 per cent of natural gas, 42 per cent of coal and 95 per cent of uranium (Commission 2014).
3. Contributions on EU climate and energy policies include: Boasson and Wettestad (2013); Birchfield and Duffield (2011); Buchan (2009); Dupont and Oberthür (2015); Dreger (2014); Jordan et al. (2010); Oberthür and Pallemaerts (2010); and Morata and Sandoval (2012).

REFERENCES

Birchfield, V.L. and J.S. Duffield (eds) (2011), *Toward a Common European Union Energy Policy: Problems, Progress, and Prospects*, New York: Palgrave Macmillan.

Boasson, E.L. and J. Wettestad (2013), *EU Climate Policy: Industry, Policy Interaction and External Environment*, Farnham: Ashgate.

Buchan, D. (2009), *Energy and Climate Change: Europe at the Crossroads*, Oxford: Oxford University Press.

Commission (2014), *European Energy Security Strategy*, COM (2014) 330 final, 28 May, Brussels: European Commission.

Costa, O. and K.E. Jørgensen (eds) (2012), *The Influence of International Institutions on the EU*, Houndmills: Palgrave Macmillan.

Di Lucia, L. and A. Kronsell (2010), 'The willing, the unwilling and the unable – explaining implementation of the EU Biofuels Directive', *Journal of European Public Policy*, 17 (4), 545–63.

Dreger, J. (2014), *The European Commission's Energy and Climate Policy – A Climate for Expertise?*, Basingstoke and New York: Palgrave Macmillan.

Dupont, C. and S. Oberthür (eds) (2015), *Decarbonization in the European Union: Internal Policies and External Strategies*, Basingstoke and New York: Palgrave Macmillan.

Dür, A., G. Mateo and D.C. Thomas (2010), 'Negotiation theory and the EU: the state of the art', *Journal of European Public Policy*, 17 (5), 613–18.

European Council (2009), *Presidency Conclusions, 29/30 October*, Brussels, 1 December 2009, accessed 13 October 2015 at http://www.consilium.europa.eu/uedocs/cms_data/docs/pressdata/en/ec/110889.pdf.

Eurostat (2007), *Energy, Transport and Environment Indicators,* 2007 edition, accessed 1 June 2015 at http://ec.europa.eu/eurostat/documents/3930297/5961574/KS-DK-07-001-EN.PDF/efe273b8-5f99-479e-8bda-9d9f1ffe809b?version=1.0.

Hooghe, L. and G. Marks (2001), *Multi-Level Governance and European Integration*, Oxford: Rowman & Littlefield.

IPCC (2014) Intergovernmental Panel on Climate Change, *Climate Change 2014, Synthesis Report*, accessed 21 April 2015 at http://www.ipcc.ch/pdf/assessment-report/ar5/syr/AR5_SYR_FINAL_SPM.pdf.

Jordan, A., D. Huitema, H. van Asselt, T. Rayner and F. Berkhout (eds) (2010), *Climate Change Policy in the European Union: Confronting the Dilemmas of Mitigation and Adaptation?*, Cambridge: Cambridge University Press.

Morata, F. and I. Sandoval (eds) (2012), *European Energy Policy: An Environmental Approach*, Cheltenham and Northampton, MA: Edward Elgar.

Moravcsik, A. (1998), *The Choice for Europe: Social Purpose and State Power from Messina to Maastricht*, London: Routledge.

Moravcsik, A. (1999), 'A new statecraft? Supranational entrepreneurs and international cooperation', *International Organization*, 53 (2), 267–396.

Oberthür, S. and M. Pallemaerts (eds) (2010), *The New Climate Policies of the European Union*, Brussels: VUB Press.

Richardson, J.J. (1996), 'Policy-making in the EU: interests, ideas and garbage cans of primeval soup', in J. Richardson (ed.), *European Union: Power and Policy-Making*, London: Routledge, pp. 3–23.
Sebenius, J.K. (1983), 'Negotiation arithmetic: adding and subtracting issues and parties', *International Organization*, 37 (2), 281–317.
Skjærseth, J.B. (2000), *North Sea Cooperation: Linking International and Domestic Pollution Control*, Manchester: Manchester University Press.
Skjærseth, J.B. and J. Wettestad (2008), *EU Emissions Trading: Initiation, Decision-Making and Implementation*, Aldershot: Ashgate.
Treib, O. (2008), 'Implementing and complying with EU governance outputs', *Living Reviews in European Governance*, 3 (5), 1–30.
Underdal, A. (1980), *The Politics of International Fisheries Management: The Case of the Northeast Atlantic*, Oslo: Universitetsforlaget.
Underdal, A. and K. Hanf (2000), *International Environmental Agreements and Domestic Politics: The Case of Acid Rain*, Aldershot: Ashgate.

2. Analytical framework

Jon Birger Skjærseth and Per Ove Eikeland

This chapter outlines the analytical framework that will guide our empirical enquiry into how and why the EU has been able to initiate, decide, implement and reform increasingly ambitious climate/energy targets and policies. Various strands of theories are combined – including negotiation theory on issue-linkages and theories of EU policymaking and implementation – to examine the different policy phases.

2.1 INITIATING AND DECIDING EU POLICIES

Unanimity has been the general rule with the adoption of EU climate/energy targets and policies. Unanimity has either been legally required in the European Council, or politically determined as in the negotiations on the 2008 climate and energy package.[1] In such cases, the policy adopted might be expected to reflect the positions held by the least ambitious actor, and the 'Law of the Least Ambitious Programme' would apply (Underdal 1980). This situation might lead to consent between actors whose agreement is necessary to change the status quo, but with low ambitiousness – particularly if new and less ambitious actors enter the negotiations.[2] And so we ask: How could the EU raise the level of ambitiousness from the status quo, without sacrificing consent among member-states whose energy-economic interests have differed considerably?

One compelling answer is that the 'Law of the Least Ambitious Programme' will apply only in the absence of issue-linkages. When several different issues are linked, the level of ambition can be raised without sacrificing consent among actors. Negotiation theory posits that issue-linking essentially involves joint negotiation of two or more issues, and generally occurs because at least one actor believes that linkage will improve the chances of a favourable outcome (Sebenius 1983; Hovi and Skodvin 2008). Combining different issues in order to achieve joint settlement can raise EU-level ambitiousness without sacrificing consent among pivotal actors whose agreement is necessary for changes in policy to be adopted (Tsebelis 2002).

There are at least three issue-linkage mechanisms that can promote EU-level unanimity departing from the status quo. First, issues that are differently valued by policymakers can be combined, creating possibilities for mutually beneficial exchange of concessions (Sebenius 1983; McKibben 2010) – as with differing concerns for energy security and climate change. Secondly, distributional obstacles can be overcome by adding issues as side-payments – a mechanism whereby 'winners' can compensate 'losers' so that everyone benefits. Direct side-payments may be institutionally difficult to arrange or insufficient for fully compensating 'losers', but issues that function as side-payments may prove effective (Tollison and Willett 1979; Sebenius 1983). For example, revenues from the auctioning of emissions-trading allowances can be used to compensate lower-income member-states for their investments in low-carbon technologies. Finally, issues can be added to exploit the interdependencies; and, with positive relationships among different issues, synergies may be exploited to the advantage of all parties (Sebenius 1983). Thus, policies for combating climate change can reduce air pollution and create new 'green' jobs, helping to make decision-makers more willing to agree on climate policies.

Two countervailing propositions tied to the consequences of issue-linkages are considered here. The first holds that adding issues can yield joint gains that create or enhance the probabilities of a successful outcome by combining differently valued issues, crafting side-payments and enabling synergies – as with the making of policy-package deals. These mechanisms can help to explain how the EU managed to adopt a new and ambitious package of climate and energy policies.

In contrast, the second view holds that adding issues may reduce the chances for successfully negotiated outcomes (Sebenius 1983). Adding issues may introduce new, pivotal actors and render policymaking more complex and cumbersome. Issue-linkage fails when individual issues have little functional commonality or when the basis for a new agreement could destroy the common ground. This second view could explain why some other, related policies either failed or were negotiated independently of the EU's core climate and energy package. In the following, these general propositions will be further specified according to the various policy phases.

Negotiations are necessary to the EU's functioning (Dür et al. 2010). However, the EU is more than international negotiations. It is a complex and decentralized multi-level political system based on the voluntary commitment of its member-states. The EU relies on these states to administer state power, but it can work as a complete political system without transforming the territorial organization of the state (Hix 2005). The EU also differs from a classic federal state such as the USA in important regards, including the separation of powers, cooperative norms and ability for long-term policymaking (Skjærseth et al. 2013). Power dispersion among the EU institutions leads to extensive consultations among the formal agenda-setter (the Commission) and the 'veto players' (the Council and the European Parliament) before the formal negotiations on legislation; and that encourages consensus-seeking when policies are initiated and negotiated. The nonpartisan Commission, led by Commissioners from each member-state, is collectively responsible for initiating new legislation, and is appointed to serve EU interests. It is able to think in a more long-term manner than most national governments that are elected by domestic constituencies for a certain period (Skjærseth 2015). Consensus-seeking and long-term policy developments are likely to promote high capacity for issue-linkage, which in turn facilitates positive outcomes.

The uniqueness of the EU has spurred several 'grand theories' of integration and mid-level theories and approaches of policymaking processes. These theories have paid little attention to negotiation theory, however.[3] Here we aim to contribute by integrating literatures on EU politics and policies to negotiation theory on issue-linkages. Negotiation theory is necessary for explaining the progress or failure of EU climate and energy policy under requirements of unanimity. Theories of EU integration and policymaking are necessary for explaining why and how policies developed in the first place.[4]

LI and MLG are two conceptual approaches commonly employed in the study of EU environmental, climate and energy policies. LI places main emphasis on the member-state level. It explains bargaining outcomes mainly as a result of the interests and preferences of these states, leaving scant room for autonomous supranational institutions to influence policymaking significantly (Moravcsik 1998, 1999). It is the states that act as 'gatekeepers' between national and international politics. The core claim of state-centric theories is that 'policy-making in the EU is determined primarily by national governments constrained by political interests nested within autonomous national areas' (Hooghe and Marks 2001, p. 3). It follows from LI that new EU policies will reflect the initiatives and constellations of interests and preferences among the member-states.

Applying this LI approach to the politics of issue-linking, we will expect the crafting of the climate and energy package to reflect the efforts of the member-states. Such initiatives can be described as instrumental or entrepreneurial leadership (Underdal 1991; Young 1991; Malnes 1995). Issue-linkages can occur when at least one (state) actor believes that linkages will improve the chances of a favourable outcome.

MLG approaches have been presented as alternatives to state-centred intergovernmentalist approaches to EU policymaking (Marks et al. 1996; Fairbrass and Jordan 2004). The many variants of MLG all share the assumption that governments are not in full control of policymaking in Brussels, and that European integration has weakened the state. In explaining EU policy, this approach brings in many different governance levels – including the EU, member-state and the subnational levels – blurring a clear separation of domestic and international politics. It follows that the development of EU policies is mainly compatible with MLG to the extent that policy has also been shaped by factors other than member-state interests, including the EU institutions and non-state actors.

Two mechanisms that 'challenge' state control are particularly relevant for this study. First, we note the role and influence of the supranational EU institutions acting as organizations and arenas for collective decision-making. The Commission can have an independent influence on policy-making above and beyond its role as agent for national governments. It can also be seen in the application of qualified majority voting and co-decision with the European Parliament. Second, there are non-state actors, such as industry and 'green' groups, that directly influence climate and energy policymaking at the EU level in Brussels, a mechanism that has gained importance as a result of the growing institutional powers of EU institutions (Hix 2005). The fundamental basis of the EU institutions is that the member-states have voluntarily transferred parts of their sovereignty to the EU. Since the 1957 Treaty of Rome, successive treaties and treaty reforms have given the EU institutions greater legislative, executive and judicial powers, culminating in the 2003 Nice Treaty and the 2009 Treaty of Lisbon (Hix 2005).

As regards the making of more ambitious climate and energy policies, we will pay particular attention to these mechanisms – especially the role of the Commission, as it has the power to initiate legislation, and non-state actors which tend to form policy 'networks' around specific interests and policies (Richardson 1996).[5] Issue-linkages and policy packages may have been crafted by the Commission with input from non-state actors in order to unify member-state preferences in favour of the Commission's agenda.

We see the LI and MLG approaches as complementary, indicating different actors, institutions and mechanisms that are all needed to explain development of EU climate and related energy policies. For analytical clarity, however, we have framed the propositions below as quite simplistic and competitive. Moreover, the LI and MLG approaches may have differing explanatory power in different policy phases. In the phase of *policy initiation*, we would expect the Commission to play a prominent role. As noted, it has the formal right to initiate and draft new EU legislation, which associates high issue-linkage capacity with the entrepreneurial role played by the Commission. However, its ability to initiate a package will depend on the degree of internal cohesion in the Commission itself, a body that has been characterized as segmented and differing in interests and culture (Sbragia 2000).[6] As issue-linkages occur because at least one actor believes the linkages will improve the chances for a favourable agreement, a Commission service with competencies over climate or energy policies may have initiated the package not only in order to unify member-state interests and preferences, but also to unify differing positions within the Commission. All this leads us to propose that *the 2008 EU climate and energy package was initiated independently (from member-states) by the Commission with input from non-state actors.* Observations pointing in this direction would strengthen MLG conceptions of EU policy initiation. In Chapter 4, we map positions and trace activities within the Commission and among non-state actors in laying the groundwork and initiating the package.

This being said, member-states can also exert significant influence on the Commission in the policy initiation phase, through advisory committees, hearings and informal consultations with the Commission (Grant et al. 2000). The Commission is appointed to serve EU interests, but the Commissioners are also a link to national publics and serve as crucial clearinghouses for the interests of the member-states vis-à-vis the Commission (Egenhofer et al. 2011). The state holding the EU presidency can also take new initiatives. Moreover, the Commission may anticipate the constellation of interests among member-states when drafting its proposals (McKibben 2010). Against this backdrop, we propose that *the Commission initiated the 2008 package at the request of the member-states.* Observations in this direction would strengthen LI ideas of EU policy initiation. In Chapter 4, we map the positions and trace the activities of some of the key member-states – collectively and individually – over time.

In *decision-making*, the member-states have a more important role than that of the Commission. The European Council, composed of the 28 heads of states and governments, has become increasingly important in developing climate and energy goals.[7] Decisions are normally taken by unanimity, but the European Council does not exercise legislative functions – its job is to define the political direction and priorities of the EU.

The Council of Ministers consists of the 28 member-states represented by relevant ministers. Its main function is to pass EU laws proposed by the Commission; here it has the final say together with the European Parliament. The member-state occupying the presidency is responsible for chairing the negotiations, and may play an important role as a mediator. The likelihood of the Commission getting the necessary backing for a proposal also depends on the decision rule applied. Unanimity gives each member-state more control but increases the probability of deadlock. The risk of deadlock can be reduced by combining negotiations on different policies and issues at the same time in 'packages' that provide ample room for mutual concessions, to the extent that complexity can be managed. With state-centred theories, we propose that *the outcome reflects the constellation of interests and preferences among the member-states.* Observations will strengthen LI to the extent that negotiations among the member-states determined the outcome. In Chapter 5, we map the positions and trace the activities of the member-states compared with the outcome and the extent to which the Commission anticipated broad-based member-state support for its proposal.

The Commission may continue to play an important role in the decision-making phase even though it is not a formal legislator (Nugent 1999). The content of what the Commission links together in its drafting of legislative proposals may have significant impacts on outcomes (McKibben 2010, p. 701). The Parliament has become increasingly important as a co-legislator together with the Council of Ministers. The co-decision procedure gives the Parliament the right to propose amendments or veto a proposal if it feels that its amendments have not been taken sufficiently into account. Non-state actors have no formal right to veto EU legislative proposals but may seek to affect decision-making through their policy networks with EU institutions. Against this backdrop, we propose that *the outcome of the 2008 negotiations reflects the Commission's formal legislative proposal and the position of the European Parliament.* Observations will strengthen MLG to the extent that the outcome reflected the Commission's proposal and negotiations between the European Parliament and the member-states. In Chapter 5, we analyse the Commission's proposal, the negotiations and the final outcome.

Common to the LI and MLG approaches is the emphasis on EU internal factors and processes. A central assumption is that changes in, and combinations of, the positions of EU institutions, non-state actors or member-states will not suffice for explaining bargaining outcomes. External energy-economic events and international regimes can deepen the explanations. Events, such as changes in energy prices or physical energy security, constitute factors that may affect the positions of pivotal decision-makers and non-state actors within the energy import-dependent EU; another is international regimes (Costa and Jørgensen 2012). International regimes can be defined as 'sets of implicit and explicit principles, norms and rules in a given area of international relations' (Krasner 1983, p. 2). Of relevance here is (changes in) the international climate regime, represented by the Kyoto Protocol and the UN Framework Convention on Climate Change (UNFCCC). The climate regime may produce commitments that put pressure on the least ambitious actors or level the playing field for European energy-intensive industries exposed to international competition. The harmonization of carbon prices or the purchase of external credits through international flexibility mechanisms can contribute to level the playing field. Conversely, failure to reach international agreement may strengthen opposition among pivotal decision-makers and non-state actors to more ambitious policies within the EU. These external factors will be explored as contextual conditions in which EU climate and energy policies develop.

2.2 IMPLEMENTATION AND REFORM

Whether the EU will succeed in achieving its aim of long-term decarbonization will depend on the *implementation* of its climate and energy policies in the member-states. Implementation usually takes place within a highly decentralized structure, which may result in a considerable gap between EU goals and actual results on the ground. Whether the EU succeeds in its long-term aims will also depend on further development and *reform* of policies and targets. Experiences with implementing policies for 2020 are likely to affect national positions on new long-term policies.[8]

Since the 1980s, the study of implementation of EU policies has developed through various phases (Treib 2008). The first phase focused mainly on the effectiveness of administrative institutions. The second phase placed implementation research within the study of Europeanization, focusing on the consequences of European integration for the member-states (Olsen 2002). The locus of research was placed at the intersection between EU and national policies, with the spotlight on the degree of compatibility between EU policy and domestic institutions. Implementation failure was seen as caused by adaptation pressure and misfit between the EU and national levels (Knill and Lenschow 2000; Knill 2001). Then, in the third phase, the role of domestic politics gained prominence.

This study of implementation in three selected EU member-states (Germany, the Netherlands and Poland) and Norway draws on insights mainly from the two recent phases. Before specifying explanations of implementation, we need to conceptualize how implementation can be assessed empirically.

Assessing Implementation

'Implementation' essentially refers here to the process of converting the EU package of climate and energy policies into domestic legislation, policies and measures, resulting in behavioural change among domestic actors that cause the problem, provide solutions, or both.

Legal transposition concerns the formal aspects of EU implementation, such as amendments to national legislation or the adoption of new laws or regulations.[9] Application concerns the adoption of new domestic policies and measures and the resultant behavioural change, like actual subsidies for increasing renewable energy and resulting increases in production.[10] The distinction between transposition and application is sometimes blurred, depending on the type of EU legislation and the activities to be regulated. Importantly, there is no automatic causal relationship between transposition and application: transposition will not necessarily lead to effective application, and national policies and behavioural change may occur also in the absence of EU-level policies.

Goal attainment is used as the main criterion for implementation. For this study, assessment of actual goal attainment is not feasible, nor is it necessarily the most pertinent criterion. It is not yet feasible, as EU policies are directed towards the year 2020 and beyond. It is not necessarily the most interesting aspect, as short-term policies form part of the longer-term vision of at least 80 per cent reduction in GHG emissions.[11] To fulfil this long-term EU vision, member-states must undertake major structural transformation of their energy systems, and will need far more ambitious policies and measures than those required for complying with the 2020 targets. Our assessment of implementation will thus include indicators showing whether the member-states are 'on track' towards the 2020 target *and* whether application appears directed towards radical long-term transformation. Alternatively, application may represent incremental adjustments to their existing policies and energy systems. In the latter case, implementation represents absorption of, or adjustment to, EU climate and energy policies adopted thus far (the 2020 package). We apply the distinction between absorption and transformation when assessing the implementation of the package in Germany, Poland, the Netherlands and Norway in Chapters 6 to 10.

Explaining Implementation

Our selection of variables is based on the MLG approach that emphasizes the relationship between the EU and the member-states. The first explanation centres on fit/misfit and adaptation pressure in the relationship between EU requirements and national status quo. The second explanation focuses on domestic politics as a complementary perspective.

Misfit and adaptation pressure

Some EU policies are more challenging to implement than others because member-states may be expected to do what they cannot do or want not to do (Di Lucia and Kronsell 2010). The 'goodness of fit' approach aims to explain 'much with little'. The basic argument is that differences between EU requirements and national institutions lead to 'misfit', which pressures member-states to agree to a change from the status quo (Knill and Lenschow 2000; Knill 2001). The 'goodness of fit' approach has been criticized for having weak explanatory power, for excluding actor interests and for being static and top-down (Treib 2008). Modified to address shortcomings, it is still regarded as a potentially powerful explanation (Di Lucia and Kronsell 2010). When this approach is applied to implementation of the EU climate and energy package, actor interests and dynamic development will be included.[12]

The 'distance' between the member-states' pre-existing policies, energy-economic situation, negotiating positions and the final EU policy outcome expresses the degree of fit. Negotiation positions as expressed in EU negotiations may vary significantly according to preference intensity – how salient various policies are for the negotiators. Pre-existing policies and the energy-economic situation will be examined to reveal underlying interests and preference intensity. Adaptation pressure can, in addition to 'distance' to EU requirements, be affected by subsequent modifications and specifications of EU legislation and by international commitments. International commitments in line with EU policies may, as noted, alleviate the pressure by levelling the playing field for actors exposed to international competition, and can lower implementation costs through international flexibility mechanisms.

The energy-economic situation is here narrowly operationalized as energy import dependency and the relative share of different fossil fuels in the energy mix compared with the EU average.[13] 'Policies' refer to the type and composition of national policies adopted before the 2008 negotiations and goals. All this leads to the following proposition: *high match between pre-existing policies, energy-economic situation, negotiation position and the final EU outcome will increase the feasibility of policy implementation.*

If policies adopted at the EU level promoted high 'fit', implementation would look bright. Conversely, if countries were pressured to accept unwanted policies, we would expect to find low fit, high adaptation pressure and opposition in the implementation phase. Because several EU policies were negotiated simultaneously as a package in 2008, issue-linkages at the EU level are likely to lead to varying fit and adaptation pressures between the different legislative package components. When unanimity is required, we expect that member-states will trade concessions between different policies according to their differing national priorities. Varying fit and adaptation pressure will in turn lead each member-state to respond to and implement the various package components differently. This expectation will be explored according to each of the package components.

Domestic politics

Also, when there is a certain level of compatibility between EU requirements and the national situation, implementation may still vary, due to variation in domestic politics over time or between different policy areas.

A simple 'model' of domestic politics explains responses to common EU policies by the state, or government itself, society and the relationship between the state and society, where domestic institutions channel influence through electoral and corporate channels. This 'model' relaxes the assumption that states are necessarily motivated to preserve the status quo and looks more deeply into the relationship between the EU and domestic political and administrative processes. New governments or administrative actors may have different preferences and bring in new priorities after EU policies have been negotiated and adopted. Societal actors may be strengthened or weakened by new EU policies, and may gain or lose political influence. As possible explanations of implementation 'failure' or 'success', we will focus on governmental and societal actors whose agreement or support is necessary to change the status quo, on policy style as linking state and societal actors in various ways, and on administrative organization that may promote or hamper implementation (Skjærseth 2000).

Governmental actors Implementation problems may arise because EU policies affect governmental actors in ways that create opposition. If such domestic actors are 'veto players', they can block or slow down the adoption of policies necessary for implementing EU requirements. By domestic 'veto players' is meant a set of specific individuals or collective actors whose agreement is necessary to achieve a change in the status quo (Tsebelis 2002). A typical veto player is a majority government.[14] Change in government from the 2008 EU negotiations to the implementation phase may thus affect the implementation of EU requirements. A new government may bring in different political priorities and cater to different constituencies. Given consent to EU policies, we propose that continuity in (majority) government over time – from EU policies to national implementation – will improve the likelihood of implementation. A majority government will make the legislative assembly less important for implementation, but competing parties may still affect the position of a government, particularly at election time.

Affected societal actors Societal actors, such as large companies or industry sectors, may support or resist higher-order policy goals. In explaining the incentives for affected societal actors to comply with policy goals, the key factor is how these sectors are affected by specific EU policies. The general assumption is that the interests of affected societal actors may deviate (more or less systematically) from governmental interests, possibly influencing the implementation of EU legislation. Determinants to be explored include the extent to which costs and benefits are widely distributed or concentrated (Wilson 1973). When costs are concentrated to specific subgroups of society while benefits are broadly distributed, there is reason to expect high incentives for opposing policies. This may lead to transposition failure but also application challenges, as captured by the idea of vertical disintegration – meaning that the aggregate of 'micro-decisions' among affected societal groups may deviate more or less systematically, or substantially, from higher-order policy goals (Underdal 1979). In some cases, segments of society may be described as social blocs characterized by tight alliances among ministries, parliamentary committees and industrial sectors (Skidmore and Hudson 1993). If such blocs see implementation of EU policies as high costs and few benefits, then implementation is likely to meet severe resistance (access to and influence on decision-making may still vary; see below). Against this backdrop, we propose that concentration of costs to specific societal actors and distribution of benefits will increase resistance to governmental policies and reduce the likelihood of implementation.

Resistance will be particularly difficult to overcome if negatively affected societal actors form social blocs. A policy-package approach will tend to increase the scope of affected sectors and may level out costs by distributing the burden between various domestic sectors: alternatively, it may reinforce cost-concentration by targeting the same sectors through different EU instruments.

Policy style and access Opposition from negatively affected societal actors does not necessarily translate into political influence. How influence is channelled between state and society may be seen as regulatory or policy styles, with policy style referring to what is assumed to be a distinct national approach to regulation. A fundamental distinction in the study of comparative environmental politics goes between policy styles that promote a high level of participation by affected societal actors and consensus-building, as opposed to imposition (see e.g. Jänicke 1992). The basic assumption is that policy styles can affect how EU policies are implemented.

As to *how* societal target groups are included, we can distinguish between *formal* and *informal* participation and access to decision-making. Some insiders with privileged status may represent the 'core', whereas others are more peripheral (Maloney et al. 1994). This may change over time and conform with, or deviate from, social blocs or segments of society. With regard to *who* is involved, an approach based on imposition can be said to be characterized by limited formal access to decision-making for negatively (or positively) affected societal actors – which in turn is likely to promote governmental policy goals. If environmental regulations entail net costs for affected actors, their strategy will often aim at watering down governmental regulations.[15] Conversely, a consensual approach is more characterized by open formal access for affected groups. That will tend to ensure more lenient governmental regulations, as the interests of the target groups will probably be reflected in domestic goals – making support for governmental policy more likely when regulations are to be implemented.[16] These points lead us to propose that *a consensual policy style (open access and broad participation) will improve the feasibility of implementation.*

A policy-package approach covering a wide range of sectors may create new alliances that affect access opportunities and enhance the probability of implementation. Conversely, a package may cement old alliances or bring together new and clashing constellations of public and private actors, which could promote new alliances that reduce the probability of implementation.

Administrative organization Due to fragmentation in administrative organization and laws, implementation problems may arise even when the affected societal actors are included and their interests are in line with governmental priorities. The need for administrative coordination is likely to be particularly high if several complex EU policies are to be implemented at the same time. Essentially, 'fragmentation' refers to the distribution of competence between governmental agencies involved in implementation. The basic assumption is that different regulatory actors will tend to perceive problems differently, and will apply differing decision-making criteria. In turn, perceptions and criteria are largely shaped by the formal roles of governmental agencies. This assumption is captured by the aphorism, 'where you stand depends on where you sit' (Allison 1971, p. 176).

Fragmentation has a horizontal as well as a vertical dimension. Horizontal fragmentation concerns the number of governmental agencies involved at each level of government, and the coordination between them. Vertical fragmentation centres on the number of levels involved: the distribution of competence among the state, regional and municipal authorities. The climate and energy package was adopted at the EU level, but implementation is to be conducted at different levels in the member-states. At the municipal level, problems may look very different from how they appear at the state level. What are seen as national 'interests' at the state level will not necessarily match preferences at the regional or municipal levels. Whenever local interests deviate from what the government has defined as the 'national interest', local opposition to national goals can be expected. Opposition may favour more, or less, stringent climate and energy policies. If local opposition favours less stringent policies and goes hand in hand with the decentralization of competence in the relevant sector, there is a risk of defection from national goals. In short, the risk of vertical and horizontal disintegration of policies between competent authorities tends to increase with increasing fragmentation of competence. Thus, we propose that *fragmentation within the state apparatus itself or between various levels of government will tend to reduce the feasibility of implementation.* A package approach may counterbalance – or deepen – fragmentation, depending on whether it is coordinated nationally and implemented coherently, or as individual pieces of legislation.

Exploring Implementation Consequences for Reform

Reform of EU policies can be analysed from the same explanatory perspectives as policymaking: LI, MLG and issue-linkages. Member-states are likely to assume a more prominent role when EU policies are to be reformed compared with when they are initiated. The reason is that costs and benefits from domestic implementation have materialized. Our explanation of reform builds on the preceding sections: experience with, and learning from, implementation of the EU climate and energy package will affect national positions on new long-term targets and policies (see Table 2.1).[17] Accordingly, new EU policies for 2030 will reflect earlier positions and experiences with implementing the climate and energy package for 2020. Support for strengthening the climate and energy package for 2030 can be expected if countries experience positive consequences from implementing the climate and energy package for 2020. These include benefits from combining concerns for energy security and climate change; side-payments to compensate sectors that stand to lose from stepping up climate policy; and synergies like innovation, employment creation and reduction of other problems such as air pollution. The absence of such benefits is expected to lead to resistance to reforms for more ambitious EU-level climate and energy policies. Resistance can be expected if new low-carbon opportunities envisaged by the EU have not been translated into domestic-level practice. Mixed experiences will most likely lead to demands for re-packing based on specific components, subject to positive experiences. Our analysis is based on implementation experiences in Germany, the Netherlands, Norway and Poland, but it should be noted that the positions taken by these four countries in the negotiations on 2030 targets and policies were representative of larger member-state groups (see Chapter 10).

Table 2.1 Summary of relationship between implementation experiences and reform

Experiences from implementation	Expected positions	Reform in the long term
Mainly negative: No or limited translation of EU-level envisaged benefits; considerable domestic implementation challenges	Opposition to strengthening more of the same policies	Stagnation
Mixed: Varying translation of EU-level envisaged benefits; moderate domestic implementation challenges	Re-packing based on positively experienced elements	Re-packing of existing issues and policies
Mainly positive: Good translation of EU-level envisaged benefits; few domestic implementation challenges	Support for strengthening more of the same policies	Strengthening of existing policies

2.3 CONCLUSIONS

This chapter has developed an analytical framework for explaining why, how and with what consequences the EU unanimously adopted a combined climate- and energy-policy package that departed significantly from the status quo. We have distinguished four phases of policymaking: initiation, decision-making, implementation and reform. Here we combine negotiation theories and theories on EU policymaking and implementation as well as EU-external factors, all of which are needed to address the research questions set out in Chapter 1. We see different theories as complementary for explaining EU climate and energy policies, but have, for analytical clarity, framed some of the most established expectations as competing. Other expectations are more explorative and complex in nature. Table 2.2 presents the core elements of the main theories and related expectations in terms of the four policy phases.

EU-external factors would be likely to affect EU climate and energy policies throughout these phases: energy-economic events and international commitments. When policies are initiated and decided, international events and commitments can place pressure on the EU or reluctant member-states and alter the playing field. When policies are implemented, international events and flexibility mechanisms can affect implementation costs. And reform of policies can reflect changes in energy-economic events and international commitments.

Initiation and decision-making will be analysed in Chapters 3 to 5, where we explain why and with what consequences some EU policies failed, became linked or were initiated and negotiated independently so as to facilitate agreement on EU climate and energy policies. Implementation and reform will be in focus in Chapters 6 to 10, as these phases must be understood in light of initiation and decision-making. We first assess and explain variation in implementation among Germany, the Netherlands, Norway and Poland and positions on new 2030 policies and beyond. Next, we analyse how these experiences and positions fed into and influenced the reform negotiations on the climate and energy framework for 2030, and examine the implications for long-term low-carbon transformation. The analysis of 2030 policies is preliminary, because the reform process has not yet been fully completed.

Table 2.2 Theories and policy phases

Theories	Initiation	Decision-making	Implementation	Reform
Issue-linkages/policy package	Unifies or separates pivotal actors favouring more ambitious package policies	Provides room for mutual concessions among pivotal actors, or impedes the common ground	Trade in concessions leads to varying 'fit', counterbalancing or reinforcing domestic politics challenges	Reflects existing policy package: strengthening, stagnation or re-packing
Liberal Intergovernmentalism	Commission initiated the package at the request of the member-states	Shaped by negotiations between member-states	n.a.[1]	Reflects member-state implementation experiences and the (new) constellation of interests and preferences
Multi-level Governance	Commission initiated the package with support from EU-level non-state actors, independently of the member- states	Shaped by Commission proposal and negotiations between the European Parliament and member-states	Relationship between the EU and member-states explains implementation with degree of 'fit' and domestic politics	Reflects implementation experiences in EU institutions, EU-level non-state actors and member-states

Note: [1] Implementation processes are 'black-boxed' by LI theories.

NOTES

1. The EU ETS was adopted by unanimity under the 'shadow' of qualified majority voting, which placed pressure on the least ambitious actors to agree.
2. This was the case with some of the ten CEECs that entered the EU from 2004. See Chapter 7 on Poland in this volume.
3. See *Journal of European Public Policy*, special issue on Negotiation theory and the EU: the state of the art, 17 (5), August 2010.
4. A broad understanding of issue-linkages resonates with the enduring grand theory of neofunctionalism for explaining European integration – that would depict the package as a result of 'spillover' from goals and policies related to one issue (such as climate) to another (such as energy). However, negotiation theories on issue-linkages include the possibility of failure and thereby skirt the deterministic nature of neo-functionalist explanations.
5. The 'policy network' concept has been defined as a *relatively stable set* of public and private actors with relationships serving as channels for communication and for the exchange of information, expertise, trust and other policy resources (Kenis and Schneider 1987, 1991).
6. As a consequence, the over 30 directorates-general (DGs) tend to put forward competing policy positions representing 'their' non-state networks that may prevent policy initiation.
7. The EU leaders agreed to the 20–20–20 targets for 2020 and the new climate and energy policy framework for 2030.
8. External events or changes in international commitments may also strengthen or reduce willingness to implement and reform EU policies.
9. Measures for implementing a directive are to be notified to the Commission. If the Commission finds that a member-state has failed to notify, it opens an infringement case.
10. Compliance is exclusively a question of adherence to legal requirements that may, or may not, involve implementation.
11. Since the ultimate goal of implementing EU climate and energy policies is to achieve a low-carbon economy, it follows that providing more behavioural change than strictly required in the desired direction would merit a higher relative 'score' on goal achievement than would providing less.
12. The three underlying assumptions are that states are motivated to preserve the status quo; that the 'fit' between EU and national levels includes the energy-economic situation; and that EU requirements may change after legislation has been adopted. EU legislation can subsequently be adjusted and specified in comitology, thus affecting fit and adaptation pressure.
13. Different fuels (coal, oil, gas) emit different amounts of CO_2 in relation to the energy they produce. Hard coal, for instance, emits almost twice as much as natural gas. See (US Energy Information Administration undated).
14. To be re-elected, lawmakers must respond to and promote affected constituency interests; hence such interests often influence policymakers' positions on issues. This in turn can create veto players, such as majority governments.
15. However, when decisions are to be implemented, resistance and opposition among target groups may become severe, since regulations are likely to be directed against their interests.
16. In a dynamic perspective, the key question is whether the national actors with responsibility for implementation were involved in shaping national positions in the EU negotiations on the climate and energy package.
17. This specific explanation can be placed within the wider 'policy feedback' literature. See Jordan and Matt (2014).

REFERENCES

Allison, G.T. (1971), *Essence of Decision: Explaining the Cuban Missile Crisis*, Boston, MA: Little, Brown.

Costa, O. and K.E. Jørgensen (eds) (2012), *The Influence of International Institutions on the EU*, Houndmills: Palgrave Macmillan.

Di Lucia, L. and A. Kronsell (2010), 'The willing, the unwilling and the unable – explaining implementation of the EU Biofuels Directive', *Journal of European Public Policy*, 17 (4), 545–63.

Dür, A., G. Mateo and D.C. Thomas (2010), 'Negotiation theory and the EU: the state of the art', *Journal of European Public Policy*, 17 (5), 613–18.

Egenhofer, C., S. Kurpas, P.M. Kaczynski and L.G. van Schaik (2011), *The Ever-Changing Union: An Introduction to the History, Institutions and Decision-Making Processes of the European Union*, 2nd edition, Brussels: Centre for European Policy Studies.

Fairbrass, J. and A. Jordan (2004), 'Multi-level governance and environmental policy', in I. Bache and M. Flinders (eds), *Multi-Level Governance*, Oxford: Oxford University Press, pp. 147–64.

Grant, W., P. Newell and D. Matthews (2000), *The Effectiveness of European Union Environmental Policy*, London: Macmillan Press.

Hix, S. (2005), *The Political System of the European Union*, New York: Palgrave.

Hooghe, L. and G. Marks (2001), *Multi-Level Governance and European Integration*, Oxford: Rowman & Littlefield.

Hovi, J. and T. Skodvin (2008), 'Which way to U.S. climate cooperation? Issue linkage versus a U.S.-based agreement', *Review of Policy Research*, 25 (2), 129–48.

Jänicke, M. (1992), 'Conditions for environmental policy success: an international comparison', in M. Jachtenfuchs and M.S. Strübel (eds), *Environmental Policy in Europe: Assessment, Challenges and Perspectives*, Baden-Baden: Nomos, pp. 71–92.

Jordan, A. and E. Matt (2014), 'Designing policies that intentionally stick: policy feedback in a changing climate', *Policy Sciences*, 47, 227–47.

Kenis, P.N. and V. Schneider (1987), 'The EC as an international corporate actor: two case studies in economic diplomacy', *European Journal of Political Research*, 15 (4), 437–57.

Kenis, P.N. and V. Schneider (1991), 'Policy networks and policy analysis: scrutinizing a new analytical toolbox', in B. Marin and R. Mayntz (eds), *Policy Networks: Empirical Evidence and Theoretical Considerations*, Boulder, CO and Frankfurt: Westview Press/Campus Verlag, pp. 25–62.

Knill, C. (2001), *The Europeanisation of National Administrations: Patterns of Institutional Change and Persistence*, Cambridge: Cambridge University Press.

Knill, C. and A. Lenschow (eds) (2000), *Implementing EU Environmental Policy: New Directions and Old Problems*, Manchester: Manchester University Press.

Krasner, S.D. (1983), 'Structural causes and regime consequences: regimes as intervening variables', in S.D. Krasner (ed.), *International Regimes*, Ithaca, NY and London: Cornell University Press.

Malnes, R. (1995), '"Leader" and "entrepreneur" in international negotiations: a conceptual analysis', *European Journal of International Relations*, 1 (1), 87–112.

Maloney, W.A., G. Jordan and A.M. McLaughlin (1994), 'Interest groups and public policy: the insider/outsider model revisited', *Journal of Public Policy*, 14 (1), 17–38.

Marks, G., L. Hooghe and K. Blank (1996), 'European integration from the 1980s: state centric vs. multi-level governance', *Journal of Common Market Studies*, 34 (3), 341–78.

McKibben, H. (2010), 'Issue characteristics, issue linkage and states' choice of bargaining strategies in the European Union', *Journal of European Public Policy*, 17 (5), 694–707.

Moravcsik, A. (1998), *The Choice for Europe: Social Purpose and State Power from Messina to Maastricht*, London: Routledge.

Moravcsik, A. (1999), 'A new statecraft? Supranational entrepreneurs and international cooperation', *International Organization*, 53 (2), 267–396.

Nugent, N. (1999), *The Government and Politics of the European Union*, London: Macmillan Press.

Olsen, J.P. (2002), 'The many faces of Europeanization', *Journal of Common Market Studies*, 40 (5), 921–52.

Richardson, J.J. (1996), 'Policy-making in the EU: interests, ideas and garbage cans of primeval soup', in J. Richardson (ed.), *European Union: Power and Policy-Making*, London: Routledge, pp. 3–23.

Sbragia, A.M. (2000), 'Environmental policy – economic constraints and external pressures', in H. Wallace and W. Wallace (eds), *Policy-Making in the European Union*, Oxford: Oxford University Press, pp. 293–316.

Sebenius, J.K. (1983), 'Negotiation arithmetic: adding and subtracting issues and parties', *International Organization*, 37 (2), 281–317.

Skidmore, D. and V.M. Hudson (eds) (1993), *The Limits of State Autonomy: Societal Groups and Policy Formation*, Oxford: Westview Press.

Skjærseth, J.B. (2000), *North Sea Cooperation: Linking International and Domestic Pollution Control*, Manchester: Manchester University Press.

Skjærseth, J.B. (2015), 'EU climate and energy policy: demanded or supplied?', in G. Bang, A. Underdal and S. Andresen (eds), *The Domestic Politics of Global Climate Change: Key Actors in International Climate Cooperation*, Cheltenham and Northampton, MA: Edward Elgar, pp. 71–94.

Skjærseth, J.B., G. Bang and M. Schreuers (2013), 'Explaining growing climate policy differences in the European Union and the United States', *Global Environmental Politics*, 13 (4), November, 61–80.

Tollison, R.D. and T.D. Willett (1979), 'An economic theory of mutually advantageous issue linkages in international negotiations', *International Organization*, 33 (4), 425–49.

Treib, O. (2008), 'Implementing and complying with EU governance outputs', *LivingRev.Euro.Gov.*, 3 (5), 1–30.

Tsebelis, G. (2002), *Veto Players: How Political Institutions Work*, Princeton, NJ: Princeton University Press.
Underdal, A. (1979), 'Issues determine politics determine policies: the case for a "rationalistic" approach to the study of foreign policy decision-making', *Cooperation and Conflict*, 14, 1–9.
Underdal, A. (1980), *The Politics of International Fisheries Management: The Case of the Northeast Atlantic*, Oslo: Universitetsforlaget.
Underdal, A. (1991), 'Solving collective problems: notes on three modes of leadership', in *Challenges of a Changing World*, Festschrift to Willy Østreng, Lysaker: The Fridtjof Nansen Institute, pp. 131–53.
US Energy Information Administration (undated), *How Much Carbon Dioxide is Produced When Different Fuels are Burned?*, accessed 14 November 2014 at http://www.eia.gov/tools/faqs/faq.cfm?id=73&t=11.
Wilson, J.Q. (1973), *Political Organizations*, New York: Basic Books.
Young, O. (1991), 'Political leadership and regime formation: on the development of institutions in international society', *International Organization*, 45 (3), Summer, 281–308.

3. Evolution of EU climate and energy policies

Per Ove Eikeland and Jon Birger Skjærseth

This chapter outlines the development of EU climate and energy policies until 2005, in terms of four periods.[1] It is intended as background for understanding the subsequent policy initiatives to be presented in Chapters 4 and 5. In the first period, the European Community sought to craft a climate and energy package of policies in order to show 'leadership by example' at the 1992 Rio Conference. The attempt was unsuccessful, but the key principles underlying today's climate and energy policies were formed in those early years. In the second period leading up to the 1997 Kyoto Protocol, the development of EU climate and energy policy stagnated. However, from 1997 climate and energy policies gained momentum in isolation from each other, due to differences in timing, policy instruments, organization and objectives. Separate policymaking resulted in conflict between energy and climate concerns in the fourth period, from the end of 2003.

3.1 EARLY 'POLICY-PACKING' EFFORTS

EU energy policy has a much longer history than climate policy – it can be traced back to the 1951 European Coal and Steel Community Treaty.[2] After oil overtook coal as the major energy source in the 1960s, the six member-states repeatedly failed to give the EC competence to formulate common responses to security-of-supply concerns (Duffield and Birchfield 2011, p. 3). National governments wanted to retain their autonomy to decide over the strategically important energy sector. Member-states rich in own oil resources, like the UK, were particularly zealous defenders of security-of-supply policy as a national competence (Eikeland 2012). The UK was to stick to this position until 2005 (see Chapter 5).

The status of energy and environmental policies was elevated with the 1986 Single European Act on realizing an internal market. This paved the way for the sector-integration principle whereby environmental concerns were to be included in other sector policies (Hancher 1990, p. 238; Padgett 1992, p. 56). Harmonization of national policies, also on taxes and environmental standards, would be required in order to eliminate barriers to trade and competition in an internal energy market (Lyons 1992, p. 7; Padgett 1992, p. 57).

In parallel to the unfolding of the internal market strategy, the gravity of the climate change problem was gaining greater international recognition. In 1985, the Commission first addressed the issue of climate change in its research policy; one year later, climate change was discussed by the European Parliament. In 1988, the first 'climate' Communication from the Commission to the Council was presented (Commission 1988). Comprehensive programmes for studying the scientific aspects of climate change were proposed. After the IPCC had been established in 1988, the Commission followed up by stressing the urgent need for a clear commitment on the part of the industrialized countries to stabilize CO_2 emissions by the year 2000. The Commission was at the helm in these early years, from problem 'diagnoses' to policy recommendations (Skjærseth 1994).

In 1990, the 12 member-states agreed on a non-binding target for stabilizing CO_2 emissions by 2000 at 1990 levels, adopted by a joint Council of energy and environment ministers. Such joint treatment of climate and energy issues shows that, from the beginning, the functional links between climate policies and energy policies were acknowledged. The European Community was importing about 49 per cent of the energy it consumed, and action on energy efficiency and renewable energy sources could promote energy security and reduce GHG emissions (Commission 1990). Energy production and energy consumption (including transportation) topped the list of sectors for policy consideration, reflecting the dominance of the energy sector as the main source of GHG emissions. With the expected growth in energy consumption and transport volumes from more trade in the new internal market, new climate policies would be needed.

The Commission services for the environment and for energy immediately started to hammer out a joint package of policy instruments to deliver on the 1990 stabilization target. DG Environment pushed for an ambitious and effective EU climate policy; DG Energy wanted security-of-supply policies that would be compatible with realization of the internal energy market. The stated ambition was for the Community to take a lead role in negotiating an international climate agreement at the

June 1992 UN Conference on Environment and Development (UNCED) in Rio de Janeiro. Climate change gave the Community an opportunity to play an international leadership role while at the same time addressing its internal energy security needs.

The Commission acted upon functional climate and energy links in its proposal for a policy package (Commission 1991). Central elements included proposals for a harmonized carbon/energy tax, a system for monitoring emissions, strengthening the existing energy-efficiency programme (SAVE) and establishing a programme on renewables (ALTENER). The need for action in the transport sector was also noted. Although the importance of concerted international action was stressed, it was held that the EC could also go it alone because the strategy could have positive internal benefits.

Unilateral action by example was to prove politically challenging. From 1991, studies conducted by the Commission showed that the costs involved would be unevenly distributed among the member-states, and most sectors of industry would be negatively affected (Skjærseth 1994). The evolving package proposal responded to the distributional challenges by including a burden-sharing element that accorded greater flexibility to countries with greater developmental needs (Ringius 1999).

The new tax proposal encountered massive opposition from the business sector, which claimed it would hurt their international competitiveness. This sparked off the most intense lobby campaign seen in Brussels until then (Skjærseth 1994; Newell and Paterson 1998). The Commission was split, with DG Taxation staunchly opposing a new Community-wide 'eco' tax. Also Commission President Delors felt that the tax should be made conditional on implementation of similar measures internationally, although he was tempted by the idea of using tax revenues to strengthen the Commission (Skjærseth 1993).

Also the Council was split, with environment ministers supporting a unilateral tax and the energy ministers tilted towards making the tax conditional on implementation of similar measures by Japan and the USA. The Dutch, German and Danish governments favoured a unilateral tax. The British government was negative to plans aired by the Commission for collecting tax revenues at the Community level to finance energy and environmental programmes, fearing that such revenues would strengthen the EC's supranational powers. Spain, Portugal and Greece were concerned about loss of competitiveness in the absence of more financial assistance from the Community. France, reliant on nuclear energy, preferred a tax based solely on carbon content.

Despite diverging views, the Commission proposed a compromise package of legislative proposals in May and June 1992. The Commission had opted for adoption before the UNCED Conference in Rio in early June, but failed to get sufficient member-state backing. Environmental Commissioner Ripa de Meana protested by not attending the Rio gathering. Some ministers, including those from Denmark, Germany and the Netherlands, accused the Commission of not going far enough, in particular by accepting the principle of tax conditionality. Others, among them the UK, claimed that the proposal was going too far, too soon (Skjærseth 1994).

The tax proposal was made conditional on the adoption of similar measures in other OECD countries, to protect energy-intensive industries exposed to international competition. The proposal for a directive on the SAVE programme was a watered-down version of what the Commission had earlier proposed. The proposal for a decision on the ALTENER programme put up a series of renewable energy targets for 2005 (with 1991 as base year), with estimates of how these would lower CO_2 emissions and contribute to the stabilization target adopted.[3] The fourth proposal was for a Council Decision to establish a GHG monitoring mechanism. In August 1992, Environment Commissioner Ripa de Meana resigned in protest, following member-state criticisms of his refusal to attend the Rio Earth Summit and disagreement with President Delors, who wanted a softer stand on the environment.

The UNFCCC adopted at UNCED in 1992 produced only vague commitments for GHG cuts for industrial countries, and no signals that other major OECD countries would implement similar taxes (Skjærseth and Wettestad 2008). The only success in terms of binding measures was the monitoring mechanism, adopted in 1993. Weakened programmes and budgets for energy efficiency (SAVE) and renewable energies (ALTENER) were adopted, none of them with EU legislative status (Oberthür and Kelly 2008). Efforts at developing a burden-sharing agreement also failed.

Thus, important building blocks for today's EU climate and energy policies were shaped nearly three decades ago: monitoring, economic climate-policy instruments, and policies for energy efficiency, renewables and burden-sharing among the member-states. The functional links between climate and energy concerns were recognized, although they were not acted upon. Neither the Commission nor the member-states managed to exploit issue-linkages politically so as to overcome distributional challenges.

The unsuccessful outcome of early climate and energy package efforts – particularly the carbon/energy tax – was determined by member-state veto under conditions of unanimity. Opposition from industry and segmentation within the Commission contributed to watering down the package proposal in ways that also served to erode the fragile member-state consensus before the package was proposed. The international climate negotiations served both as an opportunity for international leadership and as a constraining condition for internal EU action. Unilateral action on the climate/energy tax could distort the ideal of a level playing field by emplacing carbon costs on energy-intensive industries exposed to international competition.

3.2 CLIMATE- AND ENERGY-POLICY STAGNATION

After the failure to get the climate and energy 'package' adopted in 1992, EU climate and energy policies developed slowly and separately. For DG Energy, realizing the internal energy market programme dominated the policy agenda during the 1990s. Weakened versions of proposed Electricity and Gas Directives to remove barriers to trade in these markets were finally adopted in 1996 and 1998 (Eikeland 2011).[4] As part of the internal energy market programme, the Commission launched a new proposal for minimum tax levels for various energy products, this time without a CO_2 component. Six years would lapse before a highly watered-down version of this proposal was accepted by the Parliament and the Council.

In 1995, DG Energy outlined a new and broader policy agenda (Commission 1995). As the member-states were the main drivers of energy policies, three key objectives were stated for member-states to consider in energy-policy implementation: economic competitiveness, security of energy supply and environmental protection. DG Energy held that weighing the objectives would be necessary, and it was becoming increasingly evident that an internal energy market could clash with climate concerns. The main intention behind the energy market was to improve competition and bring down energy prices. However, low energy prices could conflict with climate concerns: lower prices would lead to higher fossil-fuel consumption, fewer investments in energy efficiency, and make (more expensive) renewable energy less competitive compared with coal, oil and gas. Moreover, DG Energy foresaw no immediate need for short-term security-of-supply policies, but recognized that action would be needed in the longer term to contain the EU's growing import dependence.

Regarding climate change, DG Energy stressed the need for curbing rising energy consumption and switching to non-fossil electricity – essentially nuclear and renewable electric power. The Council endorsed this agenda, adding that the clear linkage between energy policy and climatic protection would make it necessary to evaluate in depth the interrelation between environmental and energy-policy initiatives (Council 1995). In 1996, DG Energy proposed an extension of the ALTENER programme targets: to source 12 per cent of gross EU consumption from renewables by 2010 (Commission 1996). Renewable energy technologies – being indigenous resources unaffected by the price fluctuations of imported fuels – were presented as the key to future economic prosperity and competitiveness, contributing to stabilize EU energy costs (Commission 1996, p. 23). The 1996 Communication proposed differentiated goals because of major variations among member-states as regards the shares of renewables, resource potentials and policies. It also proposed a harmonized tradable certificate system based on renewable energy certificates as an energy-market compatible instrument that would apply to all utilities, be competitively neutral and cost-efficient. This idea would later re-surface in 2007 (see Chapter 4).

A follow-up 1997 Communication proposed facilitation of fair access for renewables in the electricity market, indicating a target for renewable electricity production of 23.5 per cent by 2010 – up from 14.3 per cent (Commission 1997). The Commission started drafting a renewable electricity directive, but work became seriously delayed and several draft proposals were shelved (Eikeland 2012). The most contentious aspect was the Community-level trade-based certificate system intended to replace national support mechanisms for renewable electricity. The Council and the Parliament could not agree; however, they stressed that the subsidiarity principle should prevail in the choice of support mechanism (Energy Council 1999).[5] Germany and Spain fronted the opposition to a joint Community trade-based system (Lauber 2007).

Climate-policy development became focused on the UNFCCC Treaty Conference of the Parties to be held in Berlin in 1995 (Skjærseth 2013). Prior to the Conference, the Community agreed on a mandate to negotiate strengthened international commitments based on a joint target applicable to all member-states. Reaching agreement on a common target proved difficult, however, because of disagreement on burden-sharing among the member-states. A breakthrough came in March 1997, when the Council agreed on a 15 per cent reduction target based on a 'triptych approach' that differentiated emissions for member-states in terms of three sectors: electricity production, the light domestic sector and energy-intensive industries (Ringius 1999). Different targets could then be imposed on the first two sectors, whereas identical targets across member-states could be imposed on energy-intensive industries to ensure a level playing field within the EU, in order to counter carbon leakage.

During the Kyoto Protocol negotiations in December 1997, the EU made clear it would not accept a protocol that did not include ambitious binding numerical targets. The USA was unwilling to accept such targets without the flexibility to count efforts abroad against national target achievement through various mechanisms: international emissions trading, Joint Implementation (JI) and a Clean Development Mechanism (CDM) (Grubb et al. 1999). The EU opposed such flexibility, emissions trading in particular. The compromise solution was that international flexibility would be optional. The EU got international acceptance for taking on a joint target – 'the EU bubble'. Lack of ambition from other parties, however, led the Community to lower its committed GHG emissions reduction target from 15 per cent to 8 per cent from 1990 levels by 2008–2012 – still the highest among the industrialized countries.

Thus, climate- and energy-policy initiatives developed separately and slowly from 1992 to 1997. Energy policy was driven by the member-states and focused on the internal energy market. Policy development was protracted because of diverging member-state interests, and the commitments agreed for renewables and energy efficiency were only non-binding. The development of climate policy became directed at burden-sharing and the international climate negotiations leading up to Kyoto. Ambitions of a harmonized carbon/energy tax withered due to lack of consensus. Adopting measures of a fiscal nature, such as the carbon/energy tax, still required unanimity among the member-states, even though qualified majority voting for environmental policy became the main rule with the 1992 Maastricht Treaty.

3.3 SEPARATE POLICY MOMENTUM

EU climate and energy policies gained momentum after the 1997 Kyoto Protocol. But these policies developed separately, based on differences in organization, timing, policy instruments and objectives.

In June 1998, the burden-sharing agreement was revised to reflect the EU's 8 per cent reduction target under the Kyoto Protocol, with individual commitments for the member-states ranging from +28.0 per cent to –28.0 per cent.[6] However, the 8 per cent target was to prove challenging in the absence of new EU climate policies, as emissions in the Community were on the rise – and were projected to increase further unless new measures could be taken (Oberthür and Kelly 2008). Some progress was made in transport at the initiative of DG Environment: in 1998 and 1999, the Commission signed voluntary agreements with

European, Japanese and Korean automobile manufacturer associations, aiming for a lower average emissions standard for new cars sold in the EU.

Given the significant variation in member-state climate policies, DG Environment took up the idea of the EU piloting an internal emissions-trading system based on industrial installations.[7] With the demise of the carbon/energy tax fresh in mind, DG Environment saw tradable CO_2 emission permits as an alternative, harmonized instrument compatible with the philosophy of the internal energy market (Skjærseth and Wettestad 2008). Unlike the carbon/energy tax which required unanimity (being a fiscal measure), emissions-trading could be adopted by qualified majority among the member-states and by co-decision with the European Parliament.

A March 2000 Green Paper outlined specific design issues for a mandatory cap-and-trade system (Commission 2000a). The Green Paper argued for a system based on harmonized EU-level cap-setting, with allocation of allowances based on payment by auctioning. This would be consistent with the 'polluter pays' principle, while providing opportunities for government re-cycling revenues to promote GHG abatement efforts and enabling governments to avoid taking difficult and politically delicate decisions about intercompany allocation of allowances. The system would apply to electric power producers and energy-intensive industries like cement, steel, pulp and paper, and refining.

Alongside the ETS Green Paper, the Commission launched the European Climate Change Programme (ECCP) to stimulate extended stakeholder participation in preparatory work on new climate policies. Stakeholders invited included various Commission DGs, member-state experts, industry and green groups. A series of working groups examined an extensive range of sectors and instruments – one was established to shape a common understanding of the need for an emissions trading system in the EU (EU ETS) (Skjærseth and Wettestad 2010). Only the UK, Denmark and the Netherlands were strong supporters of an EU ETS, with the remaining member-states either opposed (e.g. France and Germany) or indifferent. Most industry sectors, the green movement and the Parliament were either sceptical or opposed. By contrast, the electric power industry association Eurelectric was a firm proponent, as were the oil majors BP and Shell. The electric power industry did not face international competition beyond Europe and could compensate for a carbon price by raising the price of electricity. And the oil majors had already pioneered company-internal trading systems (Skjærseth and Skodvin 2003).

When the USA in March 2001 withdrew from the Kyoto Protocol, that opened a window for the EU to exert global climate leadership by pushing for ratification of the Protocol.[8] A forceful internal climate policy would strengthen the credibility of EU leadership. In essence, this resulted in the plans for an EU ETS evolving from an instrument for implementing Protocol commitments to an instrument for saving the Protocol itself (Skjærseth and Wettestad 2008). In October 2001, the Commission delivered its proposal for the EU ETS (Commission 2001a). The Emissions Trading Directive was formally adopted by the Parliament and the Council in 2003 – unanimously by the 15 member-states, and with only four votes against it at the final reading in the Parliament (Vis 2006). Germany, which relied on voluntary agreements with industry, had initially opposed the EU ETS, but bowed under the pressure of qualified majority voting (Skjærseth and Wettestad 2008).

The proposal and the final Emissions Trading Directive revealed two important changes from the 2000 Green Paper. The member-states did not accept a system with a cap set at the EU level and with allocation rules harmonized. The final system gave the member-states significant discretion in tailoring their own trading systems, including the total allocation of allowances, in the form of National Allocation Plans (NAPs). Secondly, the member-states did not endorse auctioning as the main principle for allocations. The final Directive instead settled on free allocation as the main allocation rule based on historic emissions. These changes weakened the environmental integrity of the system.

In July 2003, the Commission proposed an additional directive that would link the ETS to the flexible mechanisms under the Kyoto Protocol, thereby creating a direct link between internal climate policy and the international climate regime. The final text of this Linking Directive, adopted by the Council and the Parliament on 27 October 2004, extended the use of CDM/JI credits, reducing the costs of complying with the ETS. This also meant fewer incentives for industry to invest in low-carbon solutions within the EU. The use of 'cheap' external credits to comply with the EU ETS would bring down abatement costs; it was strongly favoured by power producers and energy-intensive industries alike. Ten years later, this option of using external credits would contribute to oversupply of allowances and an extremely low carbon price (see Chapter 5).

Implementation of the Emissions Trading Directive for the first pilot phase (2005–2007) started in 2003. The NAP preparation process in the member-states entailed significant challenges, as predicted. Ambitiousness was low as regards CO_2 reduction, revealed by the sum of national caps set, and there were significant delays in allocating allowances

(Skjærseth and Wettestad 2008). Nevertheless, the ETS was launched on 1 January 2005, and a price on carbon emerged (Skjærseth 2013).

Alongside the making of the EU ETS, concerns for the competitiveness of European industries were rising on the EU policy agenda, reinforced by geopolitical events and trends that had caused rapid increases in oil prices, and, hence increases in the price of gas and electricity as well.[9] The Lisbon European Council of March 2000 agreed on a new industrial growth and competitiveness strategy for the EU, declaring greater efforts to reduce barriers to trade and competition in the internal market, the internal energy market included, as an important measure.

The new higher oil prices also meant more concern about security of energy supply, not a salient issue during most of the 1990s. A November 2000 Communication prepared by DG Transport and Energy (DG TREN)[10] expressed deep regret at the lack of resources and instruments at the EU level for managing security-of-supply risks, and invited stakeholders to discuss what energy sources, solutions and instruments should be adopted (Commission 2000b). An accompanying report underscored earlier identified trade-offs between climate and energy-security concerns: the internal energy market had made the energy system more efficient – but it had also brought new challenges to security of supply by making capital-intensive investments and the development of new energy technologies more risky, and by stimulating demand by bringing energy prices down (Commission 2000c, p. 21). The report noted that the evidence of climate change would restrict security-of-supply options to those that served to reduce energy intensities, improve energy efficiencies and boost the use of clean, indigenous and renewable energy sources.

In May 2000, DG TREN finally adopted the contested legislative proposal on promoting electricity production based on renewable energy. It set the target for the share of renewable electricity to reach 22 per cent by 2010, with differential national targets being only indicative. The proposal included provisions that would make these targets mandatory in case of poor progress, and would give renewably generated power priority access to electricity grids.[11] Negotiations in the Council concentrated on the differentiated national targets, and some adjustments were made. The provision that targets would be made mandatory if progress towards the EU target proved slow was deleted, as was the provision mandating priority access to the grid for renewably generated power. A provision for mandating certification of green energy schemes to provide a guarantee that electricity sold actually came from renewables (Guarantees of Origin) was adopted only after the Commission pointed out that the provision would function to validate the electricity sold, and not as a step towards a trading instrument (Nilsson et al. 2009). The Directive promoting renewable electricity production was finally adopted in September 2001. Thus, we see that energy policymaking reflected the diversity of Council signals given in the period for pushing finalization of the internal energy market, promoting security of supply and integrating climate policies – in that order of priority.

In response to serious problems with the internal energy market due to asymmetries in implementation, in June 2003 amended Electricity and Gas Directives and two new regulations on harmonization of factors constraining access to cross-border infrastructure and trade were adopted (Eikeland 2011).[12] Also adopted as part of the internal market action programme was Regulation (EC) 1407/2002 aimed at stepping up efforts at phasing out coal subsidies, although with no clear timetable for implementation. In parallel, the Commission in 2001 proposed the target of a minimum 10 per cent interconnection capacity, to promote the internal energy market (Commission 2001b). The Council endorsed this target in 2002, but very limited money for funding infrastructure was made available.

The Council invited the Commission to propose, as soon as possible, an action plan for energy efficiency.[13] This action plan, prepared by DG TREN and presented by the Commission in March 2000, proposed directives to stimulate better energy performance in buildings and the use of combined heat and power (CHP) (Commission 2000d). The EU adopted the Energy Performance of Buildings Directive in 2002 and the CHP Directive in 2004.[14] New proposals were tabled by the Commission in late 2003 for a directive on energy end-use efficiency and energy services and an eco-design directive. These initiatives were to provide the framework for later efficiency standards for all energy-consuming products in the Community. Like the directive for the promotion of renewables in electricity production, the proposed energy-efficiency directives emplaced no legally binding quantitative energy-savings targets on the member-states.

The Commission prepared parallel policies aimed at reducing emissions from the transport sector. Particularly important was road transport, responsible for 84 per cent of total emissions in the sector (Commission 2001c). A Biofuels Directive would oblige member-states to ensure that at least 2 per cent of all transport fuels sold in 2005 would be biofuels, rising stepwise to 5.75 per cent by 2010. Also proposed were mandatory blending rates of biofuels into all road fuels from 2009. A proposal for an Energy Product Tax Directive (pending since 1997) would allow generous excise tax exemptions for biofuels, to ensure fast market uptake (ENDS Europe 2001c).

Policies to promote biofuels met support from agricultural and renewable energy associations but also considerable opposition. Environmental NGOs claimed that biofuels would put biodiversity and wildlife at risk, cause poorer air quality, and have questionable GHG mitigation effects since the production of biofuels would require input from fossil fuels (ENDS Europe 2001a). The oil industry declared the proposal 'premature', as the relevant science was still uncertain and because it gave the industry no flexibility whatsoever to meet the targets (ENDS Europe 2001c). Nevertheless, in May 2003 the Biofuels Directive (2003/30/EC) was adopted, followed by the Energy Tax Directive (2003/96/EC) in October. However, the numerous compromises required to gain the unanimous backing of the 15 EU states had trimmed the latter Directive of any near-term consequences for national tax levels (Hasselknippe and Christiansen 2003).

As we have seen, climate and energy policies gained momentum along separate tracks from the Kyoto Protocol and onwards. DG Environment initiated the world's first cross-border cap-and-trade system with support from the power industry and some major oil companies. Few member-states originally supported the trading idea. Meanwhile, DG TREN promoted the realization of the internal energy market and the reduction of energy dependency, with increasing emphasis on trade-offs with climate policy. Developing common EU policies on renewables in electricity production and transport and energy efficiency policies proved politically difficult. The targets were non-binding; moreover, the considerable potential for greater use of renewables in the heating and cooling sector remained unexploited. The development of energy policy at the EU level reflected the common view that the member-states should have sovereignty in choosing their primary energy sources and energy mix.

3.4 FROM SEPARATION TO CONFLICT

From the end of 2003, various climate and energy concerns escalated from separation to conflict, with member-state disagreements as well as deep splits emerging among the Commission services. In December 2003, Spain and the Italian Council Presidency tried – unsuccessfully – to get a text accepted in the Council Conclusions explicitly identifying the not-yet-operational Kyoto Protocol as a cost problem for European business (ENDS Europe 2003). Energy Commissioner Loyola de Palacio from Spain backed the concerns and questioned the wisdom of respecting the Kyoto Protocol if it failed to gain sufficient ratifications after the US withdrawal in 2001. To neutralize de Palacio, Commission President

Prodi, backed by Environment Commissioner Wallström, immediately declared the Commission's firm support of the Kyoto Protocol, which the EU had ratified in 2002. Wallström fiercely criticized the Spanish government and de Palacio for 'undermining' the EU's commitment to the Kyoto Protocol, stating that de Palacio should not mix personal views with those of the Commission (ENDS Europe 2004). However, de Palacio continued to challenge the EU's commitment to the Kyoto Protocol (*Financial Times* 2004). Again, Commission President Prodi had to go public, declaring the Commission's firm rejection of any calls to change its position on full implementation of the EU's 8 per cent Kyoto Protocol emissions reduction commitment. The Commission had invested significant time and energy in persuading a sufficient number of states to ratify the Protocol.

Moreover, the energy and environment Commissioners were not in line over EU post-2012 climate and energy policy development either. A new climate target was needed, as the first commitment period under the Protocol would end in 2012. DG Environment had in its environmental action plan set 2020 as a longer-term target date for policy development to reduce GHG emissions by 20–40 per cent and energy and transport sector integration as needed for target achievement. Environment Commissioner Wallström had proposed that a long-term EU renewable energy strategy should be developed with a 25 per cent target set for renewables in total energy consumption by 2020 (Environment News Service 2003). European Parliament members came out in favour of higher support for renewable energy, joining in the chorus of those who criticized Energy Commissioner de Palacio for foot-dragging when it came to setting long-term targets (Eikeland 2012).[15]

In February 2004, the French and German environment ministers supported a 20 per cent target for renewables by 2020, endorsed by a European Parliament resolution in April.[16] However, DG TREN stated that no long-term targets would be proposed before 2007 due to various obstacles, including 'technical and practical limits' on the 'cost-effective availability' of renewables (Commission 2004, p. 9). Adding to the tension was the ETS, developed by DG Environment, as this system was set to govern DG TREN's domain by affecting the costs of fossil fuels, nuclear and renewables differently. Energy ministers were also concerned about the EU ETS and possible 'market disturbances' in sectors regulated by the system (Energy Council 2005, p. 5). In March 2004, concerns for climate-policy costs reached the top EU level when the European Council requested the Commission undertake a cost–benefit analysis and consider new climate targets (European Council 2004, p. 9). Internal disagreement prevented the Commission from proposing a new post-2012 climate target.

The public quarrel between de Palacio and Wallström and procrastination in DG TREN show that, despite references to the Kyoto Protocol in energy legislation, GHG mitigation did not enjoy noteworthy priority in DG TREN in this period. The key concerns were finalizing the internal energy market and acting on the new security-of-supply agenda. The internal disagreements in the Commission were not merely irritating background noise – they actually prevented the Commission from proposing a new post-2012 climate target.

3.5 CONCLUSIONS

From the late 1980s, EU climate and energy policies developed differently, in several periods. In the first period, aided by the single European market initiative, the EU made an attempt to craft a climate and energy package of policies in order to show leadership by example at the 1992 Rio Conference. This noteworthy (albeit failed) attempt was based on the recognition of functional links between climate and energy issues. However, neither the Commission nor the member-states were able to exploit these linkages politically in a way that could overcome distributional challenges, combine different values or promote synergies.

The second period, leading up to the 1997 Kyoto Protocol, saw stagnation in the development of EU climate and energy policy. Ambitions of achieving a harmonized carbon/energy tax faded, and climate-policy development became directed at member-state burden-sharing in the international climate negotiations. Energy security was not an important concern: energy policies focused on realizing an internal energy market.

Following the adoption of the Kyoto Protocol, the third period led to new climate and energy policies developed in isolation and based on different concerns and priorities. Higher oil prices made energy security more important, and climate concerns were increasingly viewed as restricting the options for indigenous fossil-fuel supply. New directives were adopted to promote renewable energy and energy efficiency in electricity production, transport and buildings – but these policies were deficient in scope, bindingness and ambition. Climate policies followed a different path. The main policy instrument adopted was the EU ETS, which placed a price on carbon. The ETS, designed to deliver on the EU's 8 per cent commitment under the Kyoto Protocol, was actively used by the Commission as a measure to convince other states to ratify the Protocol.

During the fourth period, various policy developments, priorities and trade-offs between climate and energy concerns led to conflict from late 2003 over the costs of climate policies between DG TREN and DG Environment and some member-states. This conflict obstructed the Commission from proposing a new EU climate target for the post-Kyoto period. Linkages between climate and energy issues included trade-offs that magnified compartmentalization in the Commission, preventing more ambitious policies.

The low level of ambition evident in the energy policies adopted throughout this period is mainly in line with the tenets of Liberal Intergovernmentalism. The outcomes reflected the diversity of interests among member-states that prefer to control their energy mix and keep policy development at the national level. The European Parliament pushed for more ambitious policies on renewables and energy efficiency, but only gradually received more formal competence during this time-frame, and had scant influence in practice. There were systematic references to the Kyoto Protocol in energy legislation, but climate concerns were subordinated to energy market policies and energy-security concerns.

The development of the key climate-policy instrument – the EU ETS – is more in line with expectations that follow from Multi-level Governance. The Commission initiated the EU ETS with support from certain industries, largely independent from the member-states that either opposed or were indifferent to emissions-trading. However, the outcome also reflected the Commission's anticipation of member-state preferences for a decentralized system based on free allocation of allowances. Moreover, the ETS was unanimously agreed under the 'shadow' of qualified majority voting, which proved important for getting Germany on board. The Kyoto Protocol served an enabling function: it provided an opportunity for the EU's leadership ambitions by including a binding target that required climate policies at EU level, and by creating a direct link between the Protocol's flexibility mechanisms and the possibility of purchasing external credits to comply with the EU ETS.

NOTES

1. The European Union was formed by the Maastricht Treaty, which entered into effect in 1993. We refer to the European Community (EC) when presenting pre-1993 events and to the 'European Union' thereafter.
2. The Treaty was intended to replace wartime resource conflicts with cooperation aimed at orderly and affordable supplies of coal, which at the time covered about 80 per cent of total EC energy demand (Andoura et al. 2010, p. 7).

3. The targets were as follows: doubling the share of renewable energy in total energy demand from 4 per cent to 8 per cent; trebling the production capacity of non-hydroelectric power; and securing a 5 per cent market share for biofuels in total fuel consumption by motor vehicles.
4. Germany and France successfully fronted opposition to a speedy market opening for these sectors, whereas Britain headed the group of member-states pressing for a fast process. Britain sought to upload its national model of dismantling energy monopoly structures and creating a free-trade area in energy, but remained sceptical to a wider common EU policy.
5. The *subsidiarity principle* determines when the EU is competent to legislate, based on the principle that decisions should be taken as closely as possible to citizen level.
6. The agreement became legally binding in 2002 with EU ratification of the Kyoto Protocol – three years before the Protocol entered into force.
7. The Kyoto Protocol established emissions trading as an option between states at the international level.
8. The US withdrawal made the entry into force of the Protocol uncertain, as ratification was required by 55 parties that accounted for at least 55 per cent of 1990 CO_2 emissions from that group.
9. Brent Blend average prices jumped from USD 17.88/bbl in 1999 to USD 28.39/bbl in 2000.
10. Transport and energy services were merged (as DG TREN) in January 2000.
11. Earlier drafts had included a proposal for a harmonized trade-based EU-level support scheme. Strongly opposed by several member-states, that proposal had been removed.
12. Directive 2003/54/EC of the European Parliament and the Council of 26 June 2003 concerning common rules for the internal market in electricity and repealing Directive 96/92/EC, and Directive 2003/55/EC of the European Parliament and of the Council of 26 June 2003 concerning common rules for the internal market in natural gas and repealing Directive 98/30/EC.
13. This invitation was based on a 1998 Communication that discussed barriers to energy efficiency and possible instruments for realizing energy-efficiency potentials in various energy-use sectors, with a potential 18 per cent reduction in total energy consumption (Commission 1998).
14. Directive 2004/8/EC of the European Parliament and of the Council of 11 February 2004 on the Promotion of Cogeneration Based on a Useful Heat Demand in the Internal Energy Market and Amending Directive 92/42/EEC.
15. In autumn 2001, the Belgian EU Presidency strongly criticized de Palacio for postponing the adoption of energy-efficiency legislation (ENDS 2001b).
16. European Parliament Resolution, 1 April 2004, P5_TA-PROV(2004)0276 International Conference on Renewable Energy (Bonn, June 2004).

REFERENCES

Andoura, S., L. Hancher and M. van der Woude (2010), *Towards a European Energy Community: A Policy Proposal*, Brussels: Notre Europe, accessed 12 June 2015 at http://www.eng.notre-europe.eu/media/europeanenergy community-andoura-hancher-vanderwoude-ne-march10.pdf.

Commission (1988), *The Greenhouse Effect and the Community: Commission Work Programme Concerning the Evaluation of Policy Options to Deal with the 'Greenhouse Effect'*, COM (88) 656 final, 16 November, Brussels: European Commission.

Commission (1990), *Working Paper of the Commission on Security of Supply, the Internal Market and Energy Policy*, SEC (90) 1248, accessed 13 October 2015 at http://aei.pitt.edu/3687/1/3687.pdf.

Commission (1991), *A Community Strategy to Limit Carbon Dioxide Emissions and to Improve Energy Efficiency*, SEC (91) 1744 final, 14 October, Brussels: European Commission.
Commission (1995), *An Energy Policy for the European Union*, COM (94) 659 final, 11 January, Brussels: European Commission.
Commission (1996), *Energy for the Future: Renewable Sources of Energy, Green Paper for a Community Strategy*, COM (96) 576 final, 20 November, Brussels: European Commission.
Commission (1997), *Energy for the Future: Renewable Sources of Energy, White Paper for a Community Strategy and Action Plan*, COM (97) 599 final, 26 November, Brussels: European Commission.
Commission (1998), *Energy Efficiency in the European Community: Towards a Strategy for the Rational Use of Energy*, COM (1998) 246 final, 29 April, Brussels: European Commission.
Commission (2000a), *Green Paper on Greenhouse Gas Emissions Trading within the European Union*, COM (2000) 87 final, Brussels: European Commission.
Commission (2000b), *Towards a European Strategy for the Security of Energy Supply*, COM (2000) 769 final, 29 November, Brussels: European Commission.
Commission (2000c), *Towards a European Strategy for the Security of Energy Supply, Technical Document*, on file with author.
Commission (2000d), *Action Plan to Improve Energy Efficiency in the European Community*, COM (2000) 247 final, 26 April, Brussels: European Commission.
Commission (2001a), *Proposal for a Directive Establishing a Framework for Greenhouse Gas Emissions Trading within the European Community*, COM (2001) 581, 23 October, Brussels: European Commission.
Commission (2001b), *European Energy Infrastructure: Proposal for Decision of the European Parliament and of the Council amending Decision No 1254/96/EC laying down a series of guidelines for trans-European energy networks*, COM (2001) 775, 20 December, Brussels: European Commission.
Commission (2001c), *White Paper European Transport Policy for 2010: Time to Decide*, COM (2001) 370 final, 12 September, Brussels: European Commission.
Commission (2004), *The Share of Renewable Energy in the EU*, COM (2004) 366 final, 26 May, Brussels: European Commission.
Council (1995), Council Resolution of 23 November 1995 on the Green Paper for a European Union Energy Policy, *Official Journal*, C 327, 7 December.
Duffield, J.S. and V.L. Birchfield (2011), 'Introduction: the recent upheaval in EU energy policy', in V.L. Birchfield and J.S. Duffield (eds), *Toward a Common European Union Energy Policy: Problems, Progress and Prospects*, New York: Palgrave Macmillan, pp. 1–9.
Eikeland, P.O. (2011), 'The third internal energy market package: new power relations among member states, EU institutions and non-state actors?', *Journal of Common Market Studies*, 49 (2), 243–63.
Eikeland, P.O. (2012), *EU Energy Policy Integration: Stakeholders, Institutions and Issue-Linkages*, FNI Report 13/2012, Lysaker: Fridtjof Nansen Institute.
ENDS Europe (2001a), 'Tough road ahead for EU biofuels plan', 18 September.

ENDS Europe (2001b), 'Deleuze urges rapid EU energy-saving measures', 1 October.
ENDS Europe (2001c), 'EU biofuels support package published', 8 November.
ENDS Europe (2003), 'Competitiveness test set for EU climate policy', 15 December.
ENDS Europe (2004), 'Wallström attacks Kyoto protocol doubters', 30 January.
Energy Council (1999), *Council Conclusions on Renewable Energy Sources*, Brussels: 2176th Energy Council Meeting, 11 May.
Energy Council (2005), *Council Conclusions on Climate Change and Energy Efficiency*, Brussels: 2695th Transport, Telecommunications and Energy Council Meeting, 1 December.
Environment News Service (2003), *European Ministers Plan to Limit Climate Change*, 18 July, accessed 5 May 2015 at http://www.ens-newswire.com/ens/jul2003/2003-07-18-01.html.
European Council (2004), *Brussels European Council 25 and 26 March 2004, Presidency Conclusions*, 19 May, Brussels: European Council.
Financial Times (2004), 'Energy chief challenges EU stance on Kyoto', 26 February.
Grubb, M., C. Vrolijk and D. Brack (1999), *The Kyoto Protocol: A Guide and Assessment*, London: Royal Institute of International Affairs/Earthscan.
Hancher, L. (1990), 'Towards a free market for energy? A legal perspective', *Energy Policy*, 18, 233–45.
Hasselknippe, H. and A.C. Christiansen (2003), *Energy Taxation in Europe: Current Status – Drivers and Barriers – Future Prospects*, FNI Report 14/2003, Lysaker: Fridtjof Nansen Institute.
Lauber, V. (2007), 'The politics of European Union policy on support schemes for electricity from renewable energy sources', in L. Mez (ed.), *Green Power Markets*, Brentwood: Multi-Science Publishing, pp. 9–29.
Lyons, P.K. (1992), *EC Energy Policy: A Detailed Guide to the Community's Impact on the Energy Sector*, London: Financial Times Business Information.
Newell, P. and M. Paterson (1998), 'A climate for business: global warming, the state and capital', *Review of International Political Economy*, 5 (4), 679–703.
Nilsson, M., L.J. Nilsson and K. Ericsson (2009), 'The rise and fall of GO trading in European renewable energy policy: the role of advocacy and policy framing', *Energy Policy*, 37 (11), 4454–62.
Oberthür, S. and C.R. Kelly (2008), 'EU leadership in international climate policy: achievements and challenges', *International Spectator*, 43 (3), 35–50.
Padgett, S. (1992), 'The Single European Market: the politics of realization', *Journal of Common Market Studies*, 30 (1), 53–75.
Ringius, L. (1999), 'Differentiation, leaders and fairness: negotiating climate commitments in the European Community', *International Negotiation*, 4, 133–66.
Skjærseth, J.B. (1993), *The Climate Policy of the EC: A Study of Interests and Preferences versus EC Problem-Solving Capacity*, FNI Report 2/1993, Lysaker: Fridtjof Nansen Institute.
Skjærseth, J.B. (1994), 'The climate policy of the EC: too hot to handle', *Journal of Common Market Studies*, 32 (1), 25–45.

Skjærseth, J.B. (2013), *Unpacking the EU Climate and Energy Package: Causes, Content and Consequences*, FNI Report 2/2013, Lysaker: Fridtjof Nansen Institute.

Skjærseth, J.B. and T. Skodvin (2003), *Climate Change and the Oil Industry: Common Problem, Varying Strategies*, Manchester: Manchester University Press.

Skjærseth, J.B and J. Wettestad (2008), *EU Emissions Trading: Initiation, Decision-Making and Implementation*, Aldershot: Ashgate.

Skjærseth, J.B. and J. Wettestad (2010), 'Making the EU Emissions Trading System: the European Commission as an entrepreneurial epistemic leader', *Global Environmental Change*, 20 (2), 314–21.

Vis, P. (2006), 'Basic design options for emissions trading', in J. Delbeke (ed.), *EU Energy Law, Volume IV: EU Environmental Law, The EU Greenhouse Gas Emissions Trading Scheme*, Leuven: Claeys & Casteels, pp. 39–61.

4. Initiating the package for 2020

Per Ove Eikeland, Jon Birger Skjærseth and Lars H. Gulbrandsen

This chapter examines the initiation of the EU climate and energy package for 2020 against the backdrop of the various conflicting developments in climate and energy policy presented in Chapter 3. The initiation of new policies is analysed in the context of the theories and expectations set out in Chapter 2.

4.1 NEW MEMBER-STATES, NEW COMMISSION

On 1 May 2004, the EU was joined by eight new CEECs.[1] With their distinct energy structures based mainly on fossil fuels, and their energy import dependency on Russia, these new member-states were less attuned to the climate and energy policies already adopted by the EU (Skjærseth and Wettestad 2007). This enlargement posed new challenges for negotiating climate and energy policies at the EU level, not least because of significant changes in the composition of the Commission, the Council, the European Council and the Parliament.

In November 2004, the first Barroso Commission took office. Its incoming President, Jose Manuel Barroso, singled out the re-launch of the Lisbon Strategy as a top priority. The Commission promised to drive forward the internal market and reduce the regulatory burden on business, with assessment of all new rules for impacts on business. The Environment Commissioner, Stavros Dimas, emphasized environmental legislation aimed at boosting energy technological innovation (ENDS 2004). The new Transport and Energy Commissioner, Andris Piebalgs, received strong backing in the Parliament when he stressed the importance of integrating environmental concerns in energy policy (European Parliament 2004). Thus, the new Energy and Environment Commissioners signalled new opportunities that might bring an end to the deep conflict that had developed between their predecessors. In addition, two other Commissioners less attuned to climate and environmental concerns were to become central in the Commission's implementation of the Lisbon Agenda for growth and competitiveness: Competition Commissioner Neelie Kroes and Enterprise Commissioner Günter Verheugen.

In February 2005, the Kyoto Protocol entered into force, having achieved the requisite number of ratifications. This effectively ended the Commission dispute on whether the EU should assume the costs of climate-policy leadership *if* the Protocol did not enter into force. The same month, the Commission adopted the Communication *Winning the Battle against Global Climate Change* (Commission 2005). The analysis underlying this 2005 Communication, requested by the March 2004 European Council (European Council 2004), showed that costs and negative competitiveness effects could be minimized if project-based flexibility mechanisms linked to the Kyoto Protocol were fully used and synergies with other EU policies (like energy-security and air-quality policies) were fully exploited.

In line with the Lisbon Strategy, the Communication also placed greater emphasis on innovation whereby the EU could gain a first-mover advantage in low-carbon energy technologies that other countries would need to adopt. Technology demand pull and supply push policies, such as carbon pricing and R&D funding, would encourage technological development. This emphasis on energy technological innovation and growth opportunities could help to reduce DG Enterprise's opposition to climate policy.

However, and reflecting continued disagreement within the Commission, the 2005 Communication did not propose quantitative targets for longer-term GHG emission reductions, as had been suggested by the March 2004 European Council. It only stressed the need for an EU strategy in line with the IPCC recommendation to limit global average temperatures to no more than 2°C above pre-industrial levels, previously endorsed by the European Council (ENDS 2005a). The Communication gave assurances that the EU would continue to play a leadership role in the international negotiations on climate change; it also announced a second phase of the ECCP, instigating a review of progress in climate policies and exploring new actions for the future.

Responses were mixed. The Union of Industrial and Employers' Confederations of Europe (UNICE – now Business Europe) applauded the 2005 Communication, whereas Green groups branded it a 'laggard's proposal' (ENDS 2005b). The Council of Environment Ministers sided with the Green groups in denouncing the Commission's inability to formulate a proposal for long-term climate targets, and unanimously agreed that all developed countries should cut their GHG emissions by 15 to 30 per cent by 2020 and by 60 to 80 per cent by 2050 compared with

1990 levels (Council 2005, p. 13). The Parliament's Environment Committee strongly supported the Council's short- and long-term targets (ENDS 2005c).

At the Spring European Council in March 2005, EU leaders backed the first short-term goal but dropped the longer-term 2050 vision (European Council 2005, p. 16). The goal formulated by the EU leaders went further than the Commission's 2005 Communication, which had not proposed any specific climate targets for the EU by 2020. This shows that internal conflicts rendered the Commission incapable of acting as a climate-policy entrepreneur: it was the member-states that were pushing the process.

New energy-policy initiatives included an Energy Efficiency Communication indicating that 20 per cent energy savings could be achieved cost-effectively by 2020 by means of changes in consumer behaviour and by implementing energy-efficient technologies (Commission 2005). The subsequent Energy End-Use Efficiency and Energy Services Directive was finally agreed by the Parliament and the Council in late 2005.[2] However, the binding targets proposed by the Commission were not accepted by the Council. Concerning renewable energy policy, in early 2005 DG TREN launched stakeholder consultations on ideas for boosting biomass energy generally (including biofuels), to underpin a Biomass Plan to be adopted by the end of 2005.

In the autumn of 2005 came various initiatives directed at achieving greater coordination between climate and energy policies at the EU level. New hints of a more ambitious common EU energy policy came from the EU heads of state at their informal Hampton Court Summit in London in October (Eikeland 2012). This reflected that the cost of crude oil had nearly doubled from 2003 to 2005, with a peak in August 2005 tripling the costs (US Energy Information Administration undated). The leaders discussed a plan to improve coordination between climate policies and policies for better security of supply at the EU level. Significantly, this signalled that the UK, despite its traditional hesitance, was now prepared to accept stronger EU energy-policy integration (Eikeland 2012). From Hampton Court, member-state agreement emerged on a common response to climate change and energy-security challenges (European Council 2005; Commission 2006a).

Energy Commissioner Piebalgs utilized these new signals to undertake a full strategic review of EU energy policy, and promised a Communication outlining issues for broad stakeholder consultations. This would be more in line with DG Environment's ECCP programme. In October 2005, ECCP II was launched at a stakeholder conference attended by over 450 delegates representing all major stakeholders, including the

Commission, national experts, industry and the NGO community (Dimas 2005, p. 5). ECCP II established new working groups, including one on car emissions and one on CCS.

Next, a new platform for stakeholder consultation across various DG domains was established in December: the High-Level Group on Competitiveness, Energy and Environment. DG Enterprise hosted the group secretariat, with Commissioner Verheugen co-chairing the group with Piebalgs of DG TREN, Kroes of DG Competition and Dimas of DG Environment. Four ad hoc expert groups were to discuss the functioning of energy markets, the ETS, competitiveness of energy-intensive industries and energy efficiency. This could also serve to improve coordination between certain industry networks linked to specific DGs in support of linked climate and energy policies, such as between DG Enterprise and energy-intensive industries.

A new Impact Assessments System was also announced. The guidelines for this system included new Commission Inter-Service Steering Groups to improve coordination between Directorates-General. To oversee administrative implementation of these guidelines, in November 2005 Barroso appointed Catherine Day as new Secretary-General of the Commission. Day, who came in from the post of Director-General of DG Environment, was later to play a key role in aligning DG Environment with DG TREN. Despite these coordinating initiatives, when Barroso presented the Commission's climate and energy work programme for 2006, it proved to be little more than a long list of ongoing policy developments within the various DGs, noting the need to maintain momentum for these policy areas. No specific plans were launched for linking climate and energy legislation at the EU level.

4.2 LAYING THE FOUNDATIONS FOR THE PACKAGE

In early January 2006, external events suddenly gave impetus to security-of-supply concerns. A long-existing dispute peaked, with the Russian gas supplier Gazprom cutting off gas supplies to Ukraine's state-controlled Naftogaz.[3] As the latter continued to tap gas from the pipeline connecting Russia to the EU, reduced supply resulted. Supply was soon restored, but the event fanned distrust about Russia as a secure supplier of gas. The event was immediately picked up by several member-state governments, including the UK, to consolidate support for a more comprehensive EU energy policy (Duff 2006, p. 24).

Other international events also raised energy and climate policies higher on the EU policy agenda during 2006. International oil prices continued their steep upward trend, and Al Gore's award-winning documentary 'An Inconvenient Truth' raised public and political attention concerning climate change. The parties to the UNFCCC and the Kyoto Protocol agreed widely that new international reduction targets post-2012 should be decided by 2009 at the latest. The EU put forward its target of 15–30 per cent reductions by 2020. It also made clear its intent to expand the EU ETS globally, and used this ambition to get its member-states to submit their National Allocation Plans in time for the second trading period, 2008–2012 (Skjærseth and Wettestad 2008a). Concurrent Eurobarometer surveys showed growing public support for EU-level action on climate challenges (Eurobarometer 2008). However, a spring 2006 survey of European citizens showed that the national level was perceived as most appropriate for energy policies, especially in the new EU member-states (Eurobarometer 2006).

In March 2006, the Commission adopted the Strategic Energy Review Green Paper (Commission 2006a). It proposed a future strategy for balancing sustainable development, competitiveness and security-of-supply concerns and creating better coherence among the member-states as well as consistency in policy measures dealing with these three main energy-policy objectives (Commission 2006a). The Green Paper prepared by DG TREN marked a more open position towards linking energy and climate policy at the EU level, reflected in the fact that the trade-offs between climate and energy-security concerns that had been emphasized in earlier energy-policy papers were not directly mentioned. Strengthening EU-level energy policies in response to new energy-security challenges could be facilitated by linking energy policies to more vigorous climate policies underpinned by the Kyoto Protocol, to widespread and growing concern about climate change, and to the successful launch of the EU ETS (Skjærseth 2013).

The Strategic Energy Review Green Paper painted a dark picture of EU energy challenges stemming from poor implementation of internal energy market policy, lagging investments in infrastructure, higher import dependency on oil and gas from a limited number of suppliers, rising oil prices and a warmer climate. It proposed a 50 per cent target for low-carbon energy sources, nuclear energy included, with a deadline in 2025. A new Directive to stimulate renewables in heating and cooling was included as an option to be considered together with policies to simulate biofuels. The Green Paper also stated the need to stimulate CCS by providing economic incentives, legal certainty for the private sector and large-scale demonstration projects. CCS was linked to emissions-trading and carbon-pricing, which could 'make this a profitable option for the longer term' (Commission 2006a, p. 13).

Immediate responses from the member-states, EU institutions and non-state actors generally welcomed stronger coordination between energy and climate policies. Opinion differed on how this should be achieved and what should be prioritized. The Green movement criticized the lack of binding renewable and energy-efficiency targets and deadlines. The Confederation of European Industries welcomed greater coordination of climate and energy policy (ENDS, 2006a, b). The German Environment Minister criticized the heavy focus on energy-security issues at the expense of renewables and climate policy, as well as the fact that no targets were proposed for renewables beyond 2010 or for GHG emissions beyond 2020. The Parliament adopted a non-binding resolution stating that the Commission should aim higher than its proposed target of 20 per cent energy savings by 2020 (ENDS 2006d).

In March, the European Council concurred with the Strategic Energy Review Green Paper's description of energy challenges and the idea of a new integrated and coherent energy strategy, while underscoring that member-state sovereignty over the choice of primary energy sources and the energy mix should be respected (European Council 2006). It also backed the principle that a minimum level of the energy mix should come from low-carbon energy sources. The European Council did not adopt any specific targets but called for consideration of 15 per cent renewable energy by 2015, referring also to the 20 per cent energy-efficiency target proposed for 2020 (European Council 2006, p. 13).

The EU leaders invited the Commission to prepare a set of actions with clear timetables that could be adopted at the spring 2007 session of the European Council. While coherence, consistency and coordination at EU level were emphasized in general terms, these principles were not linked to any specific package of policies. This was also underscored by the attempt to improve coordination between non-state actors and various Commission services. The High-Level Group endorsed a comprehensive and balanced approach to energy security, sustainability and competitiveness, but its recommendations were linked to separate policy instruments and issue-areas (HLG 2006). No common timetable for policy execution was mentioned.

Following up on the Strategic Energy Review Green Paper, the Commission's plan for energy efficiency outlined its own tempo for policymaking. Policy measures were to be proposed over a six-year period from 1 January 2007 to 31 December 2012. The plan called for the use of a wide range of policy instruments, including binding minimum efficiency requirements for cars. In late October, Environment Commissioner Dimas called on his colleagues to support binding legislation on carmakers to reduce CO_2 emissions from new cars. A spokesperson for the enterprise department told the media, however, that no final decision would be taken before the results of an ongoing impact assessment were known (ENDS 2006f).

Biomass and biofuels policies followed their own path too. Discussions revealed major disagreements: on balancing the interests of large agricultural producers like France and Poland, and countries with smaller farm economies, like Denmark and the UK. Disagreements also mirrored differing interests among non-state actors. The farming and biofuels industries pressed for ambitious targets, whereas the oil and food industries and the Green groups opposed ambitious targets for various reasons, as discussed in Chapter 3. The Parliament was split. A 10 per cent target for biofuels by 2015, proposed by the Environment Committee, was rejected by the plenary (ENDS 2006e). The Parliament's plenary adopted a resolution that biofuels imports should comply with sustainability criteria as part of a comprehensive certification scheme. The compromise text called only for 'exploration' of unspecified 'measures' to guarantee sustainability. No new binding targets were agreed for biofuels. Instead, the Energy Ministers insisted that the member-states should remain free to determine their own policy approaches and to choose the sectors in which biomass would be used: heating, cooling, transport biofuels or electricity generation (ENDS 2006c).

In all these processes, policy fragmentation reflected differing interests and positions beneath the surface of general support for a new common EU energy policy and for linking with climate policy. Member-state government responses to the Strategic Energy Review Green Paper showed broad support for stronger EU-level energy policy but diverged as to the objectives to be given priority. The 'old' EU-15 member-states signalled quite broadly that mitigation of climate change topped the list, with a reformed and strengthened EU ETS as the major climate-policy instrument. The EU-15 had unanimously adopted the ETS in 2003 – before the CEECs joined the EU in 2004. The CEECs generally saw security of supply as the most pressing concern, and favoured new EU-level policies to deal with alternative channels of gas supply, better East–West energy interconnections, and keeping open any options for making use of domestic energy sources, including coal (Eikeland 2012). Poland in particular urged that energy security should dominate the other two pillars in a common EU energy policy (discussed in greater detail in Chapter 7).

The UK strongly preferred that more radical internal energy market policies should underpin all other policy initiatives – whereas France saw no need for new energy market policies. Some member-states opted for new EU-level renewable energy targets; others dismissed the idea. Germany voiced a clear demand for binding targets. Poland wanted equal support of renewables and coal, and France similar treatment of all zero-carbon energy sources. Most member-states endorsed energy-efficiency measures, but differed on whether quantitative targets should be set at the EU level. Member-states stressed that the energy mix should remain a matter of national sovereignty (Eikeland 2012).

However, the European Parliament was headed in the direction of more coherent climate and energy ambitions. A report of 28 November 2006 from the Parliament's Industry, Research and Energy Committee (ITRE Committee) for a parliamentary resolution stated that the new energy strategy should account for the radically new security-of-supply context to spur affordable energy from indigenous resources, while protecting the environment and combating climate change. The report recommended a binding target of 30 per cent CO_2 emissions reduction by 2020 and indicative 2050 targets that would comply with the objective of staying below 2°C (European Parliament 2006a).[4] The ensuing December 2006 Parliament plenary endorsed a binding climate target for 2020, but omitted any recommendations as to specific quantities (European Parliament 2006b).

Nevertheless, disagreement prevailed between and within services of the Commission on priority objectives, targets and timetables (Eikeland 2012). DG TREN's summary of the Strategic Energy Review Green Paper stakeholder consultation underscored the diversity in positions among Commission services, member-states and non-state actors. However, most stakeholder responses identified climate change as priority number one, ahead of security of energy supply (Commission 2006b).

In summary, in the first half of 2006, EU climate and energy-security concerns were raised higher on the political agenda, assisted by concerns for gas supply, oil prices and climate change. Recognition emerged – especially among member-states – on the need to strengthen and coordinate climate and energy policies at the EU level. But the various energy and climate policies followed their own paths, nested in vested networks of institutions, actors and interests. Policy diversity reflected differing interests among member-states, non-state actors and among and within the EU institutions, including the Commission.

4.3 INITIATING THE PACKAGE

The Commission had been invited by the European Council to prepare proposals for climate and energy action that could be adopted at the 2007 spring session. The challenge was formidable, as member-state interests differed widely and unanimity was required in the European Council.

In the autumn of 2006, a group of senior Commission officials, with backing from the Parliament, recognized better coordination of climate- and energy-policy initiatives as a strategy for promoting agreement within the Commission and among the member-states. The initiative came from DG Environment, which had decided in favour of a proposal for binding EU post-Kyoto targets, and now wanted to strengthen the EU ETS by an EU-level cap and to harmonize allocation procedures based on auctioning. The Commission had started working on the review of the EU ETS one year earlier.[5] Considerable impetus for ETS reform came with the Commission's November 2006 Communication on 'Building a global carbon market'.[6] One major lesson learned was the 'race to the bottom' caused by a decentralized system based on National Allocation Plans, as discussed in Chapter 3. Assessments of the NAPs had indicated that member-states had allocated overly generous allowances to protect their own industries (Skjærseth and Wettestad 2010). Verified ETS 2005 emissions in the spring of 2006 showed that 4 per cent more allowances had been distributed than were really needed. This verification of lack of scarcity in allowances led to an immediate halving in the allowance price, which then plummeted close to zero in 2007 (Skjærseth and Wettestad 2008b).[7]

Accordingly, DG Environment's first-best solution was a unilateral GHG emissions reduction goal based on a reformed ETS. However, the coal-dependent CEECs did not favour a stronger ETS; and DG Enterprise was sceptical of a unilateral climate policy, as it sought to protect the interests of the EU's energy-intensive industries. DG Environment therefore needed DG TREN as an ally to strengthen the case for more ambitious climate targets and policies within the Commission – to convince President Barroso and the college of commissioners to accept more ambitious EU climate-policy targets and instruments. DG Environment recognized that including climate policies in a package that also responded to heightened security-of-supply concerns could assist a more ambitious climate policy – notably by CCS, renewable energy and energy-efficiency policies.

DG TREN needed DG Environment to develop a common energy policy. In the consultations on the Strategic Energy Review Green Paper, a majority of respondents had indicated sustainability as the main objective of a new energy policy. Climate concerns were on the rise and provided an opportunity for DG TREN to capitalize on this momentum. A package of instruments that could stimulate the internal energy market, renewables, energy efficiency and a continuing role for indigenous fossil fuels through CCS could also satisfy various interests within the patchy DG TREN.[8] CCS would serve as 'political glue' by giving an opening for the petroleum and coal industries in climate policies. With the CCS option, the trade-off between climate and energy-security concerns could be alleviated by handling the CO_2 emissions from increasing indigenous fossil-fuel production. Heavy lobbying could be avoided by giving the fossil-fuel interests a future in a low-carbon economy.

Working within the Commission in a more cooperative atmosphere than before, a key group of individuals managed to craft and mobilize support for the 'package' idea: these people were the new heads of DG TREN and DG Environment and new senior officials able to persuade the Commissioners (Eikeland 2012; Skjærseth 2013; Dreger 2014). The Secretary-General of the Commission, Catherine Day, used her role as overall coordinator within the Commission. Coming from the post of Director-General of DG Environment, she knew and trusted the senior officers there who pushed for a package – and she had the ear of the Commission President.

President Barroso stated that fighting climate change would be among the Commission's top policy priorities for 2007, and selected energy and climate as the first key areas to be addressed in Commission policy-making. What would come out of coordinated targets and policies was uncertain. Leaks from internal discussions showed that Enterprise Commissioner Verheugen favoured the lower end of the 15–30 per cent climate target range, warning in a letter to Barroso against moving too fast (ENDS 2006g). On renewables, a target of around 20 per cent by 2020 was expected – but whether this would be binding and whether sector-specific targets would come was uncertain. Leaks in December indicated the likelihood of a 14 per cent binding target for biofuels, scrapped targets for the electricity sector and no mention of specific targets for renewable heating and cooling (Eikeland 2012). Moreover, DG Environment tried to persuade DG TREN to hold back its announced-energy efficiency plan, in order to align it with other planned policies. In October, the plan for reducing energy consumption by 20 per cent by 2020 was nevertheless presented to the Energy Council by Piebalgs, who was under pressure from the Finnish Presidency to deliver in accordance with pre-set plans (Skjærseth 2013). This indicates that DG Environment preferred having energy efficiency included in the wider climate and energy package.

Strong support from the Parliament and many governments, including those of the three biggest EU member-states, Germany, France and the UK, pointed to an ambitious GHG emission target. Both the UK and Germany had formulated impressive national climate goals. In late 2006, the British government played a particularly active role in shifting the climate-change discourse from a sole focus on the costs of mitigation towards the costs entailed in ignoring the problem. The British government-commissioned Stern Review was made public; it concluded that the benefits of ambitious, early action on climate change would far outweigh the costs of not acting. Whitehall had also opened the door to accepting a binding renewable energy target at the EU level, a point regarded as very important for Germany. As for France, it had insisted on a low-carbon energy target that would include its strong nuclear-power base.

With 2007 came a turning point for EU climate and energy policy. On 10 January, the Commission presented synchronized Communications on energy and on climate strategies prepared by DG Environment and DG TREN, respectively (Commission 2007a, b). Attending the joint press conference where the strategies were presented were Commission President Barroso and the Energy and Environment Commissioners, who gave evidence of joint Commission efforts and high political priority. The radical tone set by the two Communications was striking: 'Strong scientific evidence shows that urgent action to tackle climate change is imperative' (Commission 2007a, p. 3); 'The EU would have set the pace for a new global industrial revolution' (Commission 2007b, p. 5).

These two Communications proposed what have become known as the '20–20–20' targets: to cut GHG emissions; and increase the share of renewables and energy efficiency by 20 per cent by 2020. The proposal of reducing GHGs by at least 20 per cent compared with 1990 levels would be stepped up to 30 per cent if an adequate international climate treaty could be agreed in 2009. The twin targets can be seen as an effort to demonstrate international leadership on a post-2012 climate treaty. They also represented a compromise between DG Enterprise, which had argued for less than 20 per cent cuts, and DG Environment, which had wanted more than 20 per cent.

Both Communications underscored the *synergies* between climate and energy policies: an ambitious climate policy would contribute to the realization of energy goals; an ambitious energy policy would contribute to the realization of climate-policy goals. Action on climate change was placed at the centre of new EU energy policies that would also improve security of supply by stimulating indigenous renewable energy and energy efficiency, reducing the need for imported hydrocarbons. Competitiveness would benefit because exposure to fluctuating oil and gas prices would be reduced. New policies to complete the internal energy market would contribute to more effective application of the ETS and make the EU more competitive. New policies to stimulate renewables, energy efficiency and energy technology development would create new 'green' growth and jobs, and thus be compatible with the Lisbon Agenda. And CCS could allow for continued production of indigenous fossil fuels.

Largely swept under the carpet were several potential trade-offs. These included lower incentives for alternative energy investments as a result of the internal energy market's downward pressure on energy prices; potential conflicts of downward pressures in carbon prices caused by renewables and energy-efficiency measures in ETS sectors; carbon leakage; competition between land use for food production, energy and biodiversity; and more energy-intensive production of materials to improve insulation in buildings. Differences between the two 2007 Communications reflected disagreement on biofuels and sustainability: the Energy Communication (2007b) included a proposal for a 10 per cent binding share for biofuels by 2020, not mentioned in the Communication prepared by DG Environment (2007a). Sector targets for electricity, heating and cooling had been dropped.

To deliver on the GHG reduction targets, the Climate Policy Communication referred to a new action plan that included strengthening the EU ETS, limiting transport emissions by including aviation in the ETS, and reducing CO_2 emissions from cars and transport fuels. Also included was a list of instruments for which other sector services had lead responsibility, with specific cross-reference to the Energy Policy Communication.

The Energy Policy Communication set up a broad ten-point action plan for new European energy policy, essentially reflecting earlier energy priorities but now with more emphasis on climate policy. Policies to complete the internal energy market topped the list, with a long list of concrete proposals. Number two on the list was solidarity between the member-states in connection with stocks of gas and oil for a common security-of-supply policy. Third was a cross-reference to DG Environment's plans of revising the EU ETS as the basis for a long-term commitment to GHG reduction. The next points on the list outlined energy-efficiency policies (referring to the already proposed action plan) and renewable energy policies.

Also on the action list was a new European Strategic Energy Technology Plan to promote better coordination of R&D as a way of lowering the cost of clean energy and putting EU industry at the forefront of the rapidly growing low-carbon technology sector. The Plan would assist in achieving the 2020 targets; for 2030, it would focus on the commercialization of CCS, second-generation biofuels and hydrogen. And for 2050, the Plan envisioned technological development for complete decarbonization of the EU economy.

CCS policies were given special attention in a separate Communication, aimed at achieving zero emissions from coal after 2020 (Commission 2006a). This Communication was important for getting the fossil-fuel industry and the coal-based CEECs on board by recognizing the importance of fossil fuels, not least the contribution of coal to security of energy supply compatible with climate policy. The Commission proposed a substantial increase in R&D funding, including the construction and operation by 2015 of up to 12 large-scale CCS demonstration plants in commercial power generation. This would require steps to make new fossil-fuel power plants equipped with 'capture-ready' technology, to assess the potential risks from CCS, and to set requirements for licensing and risk management to remove uncertainties for investors.

An associated impact assessment analysed the impact of climate policies on renewables – not the other way around, like the impact of the renewable targets on carbon prices (Commission 2007b, p. 15). The EU ETS as such was deemed inadequate for achieving the 20 per cent renewable target, but further progress could be achieved by strengthening the system.[9]

In late January 2007, DG Environment adopted its proposal for a revised Fuel Quality Directive (FQD). The main change concerned CO_2 emissions standards for fuels marketed in Europe. Specifically, fuel suppliers (oil and refining companies) would be required to monitor and cut life-cycle carbon emissions from their marketed fuels by 1 per cent annually in the period 2010–2020, in total a 10 per cent cut (ENDS 2007).[10] The FQD was linked to the emerging package by the inclusion of a CO_2 component and the 2020 target date. Politically, it was de-linked from the other policies, as the timing of the proposal for a binding Directive differed. Other policies that were to become the core climate and energy package would be translated into formal proposals a year later.

In early February 2007 the Commission presented its delayed and revised CO_2 and Cars Strategy (Commission 2007c). The strategy expected a launch date for legislative proposals of late 2007 and included a set of measures aimed at influencing both the supply and the demand side of the EU market for cars and vans.[11] It envisaged binding EU legislation replacing the voluntary agreements to ensure that average emissions from new cars sold in the EU-27 would reach the 120 g CO_2/km target by 2012 (Gulbrandsen and Christensen 2014). Like the FQD, the car proposal became linked to the package by the inclusion of a 2020 target date. Politically, it was de-linked from the other proposals because its timing was different.

4.4 ADOPTION OF THE 20–20–20 TARGETS

The emerging energy and climate package met mixed reactions. In February 2007, a Parliament resolution on climate change upheld the recent energy-policy resolution call for a unilateral 30 per cent GHG reduction target, a 25 per cent renewable energy target and 12.5 per cent target for biofuels (European Parliament 2007). The resolution urged the Commission to propose sector-specific targets for electricity generation, heating and cooling; to introduce sustainability standards and a certification scheme for biofuels; and to impose a binding target of 120 g CO_2/km for new passenger cars marketed in the EU in 2012.

Green groups demanded 30 per cent unilateral GHG cuts, whereas energy-intensive industries expressed concerns about competitiveness if the EU should commit to a GHG target ahead of other major emitting countries. Business Europe opted for only a *non-binding* renewable energy target, urged a target of 40 per cent for nuclear electricity, and called for special measures to protect energy-intensive consumers from the rise in energy prices that was expected to follow from investments in renewables. The renewables industry lobby EREC (European Renewable Energy Council) opposed scrapping the sector targets for renewables (EREC 2007). The oil-industry lobby group Europia was infuriated with the FQD proposal, and held that the greatest scope for CO_2 reduction lay in car technology (EurActiv 2007a). The European Automobile Manufacturers Association (ACEA) denounced the new car regulation target as arbitrary and too stringent (Christensen and Gulbrandsen 2012).

Despite mixed reactions, in March 2007 the European Council adopted the key elements of the new linked climate and energy policies and targets. The EU leaders agreed that 20 per cent GHG emissions reduction and the 20 per cent increase in renewable energy consumption (including the 10 per cent biofuels targets) should be made binding. Also adopted was a non-binding commitment to reduce EU energy consumption by 20 per cent through energy-efficiency investments.

As could be expected, the European Council placed more emphasis on burden-sharing between the members than did the Commission. It concluded:

> a differentiated approach to the contributions of the Member States is needed reflecting fairness and transparency as well as taking into account national circumstances … It recognizes that implementation of these targets will be based on Community policies and on an agreed internal burden-sharing and invites the Commission, in close cooperation with the Member States, immediately to start a technical analysis of criteria. (European Council 2007, p. 12)

The European Council further stressed that differentiated national targets should be set for GHG emissions reductions and total renewables. The latter should take into account differing national circumstances, starting points and potentials. The bindingness of the specific biofuels target was made subject to production being sustainable and second-generation biofuels being available. The European Council also endorsed a review of the EU ETS as the key climate-policy instrument and agreed to go ahead with new internal energy market policies and to set up a European Strategic Energy Technology Plan. Further, the European Council welcomed the construction and operation by 2015 of up to 12 CCS demonstration plants.

The European Council agreement masked differences of opinion among EU leaders on the target for renewables. France and Finland argued for a broader low-carbon target that would include nuclear energy. Many of the new member-states, including Poland, Hungary and the Czech Republic, were heavily reliant on coal and thus sceptical of targets that could force them to invest in more expensive renewable energy sources. With the support of Britain, Germany's Chancellor Angela Merkel argued strongly for a binding target (EurActiv 2007b). Poland, Hungary and the Czech Republic also voiced concern on how efforts to achieve the climate target would be shared. Despite these differing opinions and concerns, the Commission had now gained backing from the highest political level and the two main veto players or policymaking bodies in the EU: the European Council and the Parliament.

The Commission was invited to come forward with more specific legislative proposals based on the European Council's conclusions. DG Environment and DG TREN cooperated closely throughout 2007. This was based on joint modelling work to prepare for impact assessment and strengthen burden-sharing to ensure political feasibility. As the EU ETS already covered nearly half of total EU emissions from energy-intensive industry and power production, national differentiation would have to be based on non-ETS sectors like transport, buildings, agriculture and waste.

The FQD and Car Emissions Regulation had at this stage been politically decoupled from the package. Also energy-efficiency policies continued to follow a separate track, in accordance with previous plans. The choice of not including in the equation differentiated national targets to attain the EU 20 per cent energy-efficiency goal was due to lack of political willingness, combined with separate policy dynamics and technical challenges. As noted, DG Environment tried unsuccessfully to coordinate with the energy-efficiency strategy that followed a separate timeline in DG TREN. That the member-states had never shown any actual will for binding national energy-efficiency targets was reflected by the fact that the 20 per cent EU target was not binding. Moreover, energy-efficiency policies could have burdened the CEECs, which faced the greatest energy-efficiency challenges – or, put differently, had the highest energy-efficiency potential. Negotiations would have been even more complex had energy efficiency been added to the climate and renewable energy targets. Moreover, measuring energy efficiency was challenging, and model technicalities would have made the modelling exercise more complex.

High hopes for improving compatibility between new renewable policies and the internal energy market were dashed in the course of 2007. A draft of the new Renewable Energy Directive was based on a mandatory and common trade-based certificate system, to make it compatible with the internal energy market policies for which a separate policy package was adopted by the Commission in September 2007 (Nilsson et al. 2009; Eikeland 2011; Dreger 2014). A group of member-states preferred national freedom to choose, and lobbied intensely to get this mandatory system removed; other member-states supported it. With little common ground for agreement, the Commission withdrew the proposal and made the trading instrument merely optional, which meant political de-linking of the internal energy market package from the climate and energy package. That removed one clear obstacle to embarking on further negotiations on the package. On the other hand, differentiated national renewable energy policies would continue as a barrier to electricity trade in the internal energy market.

4.5 CONCLUSIONS

The Commission initiated a package of ambitious climate and energy policies based on two issue-linkage mechanisms. First, issues that were differently valued among individual and collective policymakers as regards energy security and climate mitigation were combined to be advantageous to most policymakers. This combination helped to unify support from pivotal actors within the Commission and among the member-states. Linking CCS to the package proved politically decisive for successful initiation in two ways – first, by easing the functional trade-off between indigenous fossil-fuel-based energy security and climate concerns, and by reducing opposition from fossil-fuel interests, including powerful oil companies and coal-based member-states. Second, there was considerable emphasis on complementarities and synergies between climate and energy policies, as well as between these policies and other policies. Synergies were to be created by climate and energy goals that were mutually reinforcing, so as to reduce air pollution, create green jobs and stimulate technological innovation. Trade-offs were largely ignored.

Other related issues and policies were de-linked from the emerging package. Including energy efficiency in the package would have made the modelling exercise more complex and could have magnified differences in interests, particularly between the CEECs and the EU-15. Moreover, energy-efficiency policies followed their own institutional path. Also road transport policies were decoupled from the package, since they would serve to increase the regulatory burden for only a few member-states. The Car Emissions Regulation would affect car-manufacturing countries, and the FQD would affect petroleum-producing countries. Finally, a stronger link between the Renewable Energy Directive and the internal energy market failed. The issue-linkages and outcomes in the initiation phase are generally in line with the expectations put forward in Chapter 2 on how the linking and decoupling of issues and policies can facilitate agreement.

Multi-level Governance and the Liberal Intergovernmentalist approaches both help to explain different aspects of the emerging package. In line with LI expectations, the EU member-states, collectively and individually, changed the Commission's room for manoeuvre by encouraging it to initiate more coherent EU climate and energy policies. Thus, they put pressure on the Commission, which was initially unable to act forcefully on its own because of internal rivalry. The UK's Hampton Court initiative was particularly important, but no single member-state can be said to have been an 'entrepreneurial leader' in initiating the package. The member-states invited rather than requested the Commission to initiate policies based on linking climate and energy policies, without indicating which form the package should take.

In line with MLG expectations, it was the Commission, specifically DG Environment, that took the initiative regarding the package in 2006, and allied with DG TREN in order to strengthen the Commission's climate-policy proposal. This initiative was not an independent move: it developed as a result of interaction with the member-states. After a history of conflicts between these two DGs, changes in key personnel made joint effort on the package possible. Different non-state actor networks were linked to different Commission services and did not align in support for linking EU climate and energy policies. But some efforts were made to improve the alignment of positions, such as the High-Level Group. Finally, the package was initiated also as a response to external energy events and international commitments – specifically the rise in oil prices, the Ukraine/Russia dispute over gas supplies and the entry into force of the Kyoto Protocol. The Kyoto Protocol provided an opportunity to save costs by purchasing external credits, as well as making it possible for the EU to show leadership in the international negotiations on a new post-Kyoto treaty.

NOTES

1. Cyprus and Malta also became EU members at this time.
2. It requested member-states to achieve 1 per cent annual energy savings in the supply and distribution of energy, including transport fuels, in the period 2008 to 2016.
3. The dispute over gas supplies, prices and debts involved politicians from both countries. When Russia cut off gas supplies to Ukraine, the latter answered by diverting gas bound to the EU from transit pipelines crossing Ukrainian territory.
4. The report from the Parliament also backed the extension and revision of the EU ETS, with full auctioning of allowances and allocation of revenues from auctioning into energy R&D as the instrument for reaching the target. The Parliament further called for at least 20 per cent energy-efficiency improvements and 25 per cent share of renewables by 2020, with binding sector targets for renewables. An indicative target for renewables of 50 per cent by 2040 should be adopted. The report called for CCS in relation to coal and gas and oil.
5. The revision of the EU ETS was based on Article 30 of the 2003 Directive, which stated that, on the basis of experiences and progress achieved, the Commission was to draw up a review report by 30 June 2006, accompanied by proposals as appropriate.
6. A special working group under ECCP II was to prepare recommendations for a revised ETS in close collaboration with stakeholders. The ECCP II Working Group on the review of the EU ETS met four times in the spring of 2007. Ensuing protests by industry resulted in two ad hoc consultation meetings in 2008 with stakeholders within the ECCP II.
7. In addition to the lack of scarcity of allowances, the price collapse was caused by the regulation against transferring, or 'banking', allowances to the second phase. The Commission intervened in the second round of allocation of allowances in 2006 and forced member-states to adopt less generous NAPs for the second phase.
8. A strengthening of the ETS would be compatible with the Internal Energy Market Directorate; a more active policy on CCS would extend the future for fossil fuels and be in line with the interests of the Energy Directorate; policies on renewables and energy efficiency corresponded to the directorate responsible for these policies.
9. Still, a single CO_2-based target achieved through the ETS alone could be an appropriate approach, if the renewable energy sector were a mature and well-functioning market (Commission 2007d, p. 28).
10. Fuel suppliers could do this by cutting fuel-processing emissions or by a higher share of lower-carbon fuels. To facilitate the market uptake of biofuels, the Communication proposed allowing a higher blending of biofuels with ordinary fuels (to raise the standard from 5 to 10 per cent). The proposed directive would also force the petroleum industry to calculate the life-cycle carbon balance of biofuels to ensure a real GHG emissions reduction effect.
11. An 'integrated approach' meant that improvements in motor technology would have to reduce average emissions to no more than 130 g CO_2/km; 10g would be left to softer 'complementary' measures: use of biofuels, fuel-efficient tyres and air-conditioning, traffic and road-safety management and changes in driver behaviour (Commission 2007d). The strategy did not contain any binding longer-term target, but a 95 g CO_2/km by 2020 was envisaged through accelerated investments in research, through member-state stimulation of market uptake (via labelling and car taxation), and by instructing car manufacturers to sign an EU code of good practice on car marketing and advertising.

REFERENCES

Christensen, A.R. and L.H. Gulbrandsen (2012), *EU Policies on Car Emissions and Fuel Quality: Reducing the Climate Impact from Road Transport*, FNI Report 14/2012, Lysaker: Fridtjof Nansen Institute.

Commission (2005), *Winning the Battle against Global Climate Change*, SEC (2005) 180, 9 February, Brussels: European Commission.

Commission (2006a), *Green Paper on a European Strategy for Sustainable, Competitive and Secure Energy*, COM (2006) 105 final, 8 March, Brussels: European Commission.

Commission (2006b), *Summary Report on the Analysis of the Debate on the Green Paper 'A European Strategy for Sustainable, Competitive and Secure Energy'*, Commission staff working document, SEC (2006) 1500, 16 November, Brussels: European Commission.

Commission (2007a), *Limiting Global Climate Change to 2 Degrees Celsius: The Way Ahead for 2020 and Beyond*, COM (2007) 2 final, 10 January, Brussels: European Commission.

Commission (2007b), *An Energy Policy for Europe*, COM (2007) 1 final, 10 January, Brussels: European Commission.

Commission (2007c), *Results of the Review of the Community Strategy to Reduce CO_2 Emissions from Passenger Cars and Light-Commercial Vehicles*, COM (2007) 19 final, 7 February, Brussels: European Commission.

Council (2005), *Council Conclusions Environment*, 2647th Council Meeting, 10 March, Brussels.

Dimas, S. (2005), *Developing the European Climate Change Programme*, speech at ECCP stakeholder conference, 24 October, Brussels, accessed 13 October 2015 at http://ec.europa.eu/clima/events/docs/0003/dimas_en.pdf.

Dreger, J. (2014), *The European Commission's Energy and Climate Policy: A Climate for Expertise?*, Basingstoke and New York: Palgrave Macmillan.

Duff, A. (2006), *Plan B: How to Rescue the European Constitution*, Notre Europe Studies and Research No. 52, Brussels: Notre Europe.

Eikeland, P.O. (2011), 'The third internal energy market package: new power relations among member states, EU institutions and non-state actors?', *Journal of Common Market Studies*, 49 (2), 243–63.

Eikeland, P.O. (2012), *EU Energy Policy Integration: Stakeholders, Institutions and Issue-Linking*, FNI Report 13/2012, Lysaker: Fridtjof Nansen Institute.

ENDS (2004), 'Dimas faces up to a sceptical parliament', 29 September.

ENDS (2005a), 'EU shows caution on future climate targets', 9 February.

ENDS (2005b), 'EU argues over tactics for post-Kyoto talks', 28 February.

ENDS (2005c), 'MEPs urge stronger EU climate policies', 12 October.

ENDS (2006a), 'EU energy paper draws praise and criticism', 8 March.

ENDS (2006b), 'Germany's Gabriel slams EU energy paper', 16 March.

ENDS (2006c), 'Council strikes a balance on biomass energy', 12 May.

ENDS (2006d), 'Climate top of next year's agenda, says Dimas', 21 June.

ENDS (2006e), 'MEPs demand EU ban on palm oil biofuels', 4 October.

ENDS (2006f), 'Dimas pushes for EU car CO2 emissions law', 30 October.

ENDS (2006g), 'EU must get real on energy-climate policies', 24 November.

ENDS (2007), 'EU fuel quality law revision proposed', 31 January.
EREC (2007), *EREC Position on the Need for Sectorial Targets*, Brussels: EREC.
EurActiv (2007a), 'EU aims at oil firms on way to low-carbon future', 1 February.
EurActiv (2007b), 'EU makes bold climate and renewables commitment', 9 March.
Eurobarometer (2006), Europa, Press Release 2006. European citizens in favour of a European Energy policy, says Eurobarometer survey, IP/06/66, Brussels, 24 January.
Eurobarometer (2008), *Europeans' Attitudes towards Climate Change*, Brussels: European Commission.
European Council (2004), Conclusions from European Council 25 and 26 March 2004, Brussels, 19 May.
European Council (2005), Presidency Conclusions from European Council 22 and 23 March 2005, Brussels, 23 March.
European Council (2006), Presidency Conclusions from European Council 23 and 24 March 2006, Brussels, 18 May.
European Council (2007), Presidency Conclusions from European Council 8 and 9 March 2007, Brussels, 2 May.
European Parliament (2004), Committee on Industry, Research and Energy hearing with Mr Andris Piebalgs, ITRE/2004/D/49530, Brussels, 15 November.
European Parliament (2006a), *Report on a European Strategy for Sustainable, Competitive and Secure Energy: Green Paper* (2006/2113(INI)), Committee on Industry, Research and Energy, Rapporteur: Eluned Morgan, FINAL A6-0426/2006, 28 November.
European Parliament (2006b), Resolution on a European Strategy for Sustainable, Competitive and Secure Energy: Green Paper, Strasbourg, 14 December.
European Parliament (2007), European Parliament Resolution on Climate Change, P6_TA (2007) 0038, Brussels, 14 February.
Gulbrandsen, L.H. and A.R. Christensen (2014), 'EU legislation to reduce carbon emissions from cars: intergovernmental or supranational policy-making?', *Review of Policy Research*, 31 (6), 503–28.
HLG (2006), *Contributing to an Integrated Approach on Competitiveness, Energy and Environmental Policies*, Brussels: HLG, 2 June.
Nilsson, M., L.J. Nilsson and K. Ericsson (2009), 'The rise and fall of GO trading in European renewable energy policy: the role of advocacy and policy framing', *Energy Policy*, 37 (11), 4454–62.
Skjærseth, J.B. (2013), *Unpacking the EU Climate and Energy Package: Causes, Content and Consequences*, FNI Report 2/2013, Lysaker: Fridtjof Nansen Institute.
Skjærseth, J.B. and J. Wettestad (2007), 'Is EU enlargement bad for environmental policy? Confronting gloomy expectations with evidence', *International Environmental Agreements: Politics, Law and Economics*, 7, 263–80.
Skjærseth, J.B and J. Wettestad (2008a), *EU Emissions Trading: Initiation, Decision-Making and Implementation*, Aldershot: Ashgate.

Skjærseth, J.B and J. Wettestad (2008b), 'Implementing EU emissions trading: success or failure?', *International Environmental Agreements*, 8, 275–90.
Skjærseth, J.B. and J. Wettestad (2010), 'Fixing the EU Emissions Trading System?
Understanding the post-2012 changes', *Global Environmental Politics*, 10 (4), 101–23.
US Energy Information Administration (undated), *Europe Brent Spot Price FOB*, accessed 3 June 2015 at http://www.eia.gov/dnav/pet/hist/LeafHandler.ashx?n=pet&s=rbrte&f=a.

5. Deciding the package for 2020

Jon Birger Skjærseth, Per Ove Eikeland and Lars H. Gulbrandsen

How and why were new, binding EU climate and energy policies linked in a specific policy package and agreed in 2008? Also in this chapter, our empirical examination of the negotiations and decision-making is guided by the analytical framework set out in Chapter 2.

5.1 THE PACKAGE PROPOSAL

In January 2008, the Commission formally proposed the climate and energy package for achieving the 20–20–20 targets by 2020. The main structure involved two cross-sector climate-policy instruments prepared by DG Environment. The first was a revised EU ETS aimed at reducing emissions in the ETS sectors – the energy-intensive and power-producing industries. The second instrument was an ESD based on national targets for sectors not covered by the ETS, such as transport, waste, buildings and agriculture. These two instruments covered most of the EU's GHG emissions. The 20 per cent GHG reduction target by 1990 translated into a 14 per cent reduction compared with 2005 levels, which were used as the baseline. The EU ETS sectors were to achieve a 21 per cent reduction and the non-ETS sectors a 10 per cent reduction, compared with 2005.

Two specific low-carbon directives were also proposed in the 'core' package: the RED prepared by DG TREN and the legal framework for safe CCS prepared by DG Environment. The ETS, ESD and RED proposals were linked through common impact assessment on cost distribution for the member-states and industry. Other functionally related policies were associated with the 'core' package by a shared 2020 deadline, but were not made subject to common impact assessment and followed different time schedules for negotiation and adoption. These policies included energy-efficiency and road-transport legislation for reducing emissions from fuels and new cars.

The ETS reform proposal was the key climate-policy component of the package. Significant changes were evident compared with the 2003 Emissions Trading Directive. First, the Commission proposed introducing a single, EU-wide cap and allocating allowances on the basis of fully harmonized rules for the period 2013 to 2020. That meant that NAPs would no longer be needed. The linear annual reduction in the ETS sectors consistent with the '21 per cent below 2005 emissions by 2020' target amounted to 1.74 per cent per year, specified in the ETS cap.

Second, auctioning was now proposed as the main principle for all allocation of allowances. For the electric power industry, which was not exposed to competition beyond Europe due to limited transmission capacity, full auctioning was proposed from 2013 onwards. Auctioning revenues should be used to redistribute revenues to lower-income member-states, and as a general mandatory instrument for all member-states to finance measures for climate mitigation and adaptation.

For other industries, a transitional period was foreseen, during which the amount of free allocations would be gradually reduced from 2013, reaching zero by 2020. In the event that other developed countries and other major emitters did not join in an international climate agreement, energy-intensive sectors would, according to the Commission, probably be subject to carbon leakage. The Commission proposed to delay identifying such carbon-leakage-prone sectors until after the December 2009 Copenhagen Summit, promising a report on the subject by 30 June 2010. The Commission indicated that energy-intensive industries exposed to significant risk of carbon leakage could receive up to 100 per cent of allowances free of charge (Commission 2008a, p. 8).

Third, access to CDM credits and free allowances through benchmarking were intended to serve as effective strategies for limiting the potential negative effects of the revised ETS as regards competitiveness (Commission 2008b). However, no new CDM or JI credits were to enter the system (only those banked from the 2008–2012 phase) – unless agreement could be reached on a 'satisfactory' new global climate agreement in Copenhagen, and the EU moved to the more ambitious 30 per cent GHG reduction goal. In that case, CDM and JI credits would cover up to half of the additional reduction effort needed. This move directly linked the ETS reform proposal to the outcome of the international climate negotiations.

The Effort-Sharing Decision was developed in the context of the earlier burden-sharing arrangement related to the Kyoto Protocol (see Chapter 3). The change in wording from 'burden-' to 'effort-' sharing signalled a more positive and opportunity-based approach. The division of the burden of reductions between the ETS (21 per cent reduction) and non-ETS (10 per cent reduction) sectors was based on a cost–efficiency scenario developed by the Commission: more reduction would be required in the ETS sectors because abatement costs were lower there (particularly in the electricity sector) than in most other sectors.

The European Council meeting in March 2007 had not introduced a clear distinction between the ETS and the non-ETS sectors or mentioned a specific ESD instrument, but emphasized fairness and differing national circumstances. The Commission faced a difficult situation with ten new CEEC members (Romania and Bulgaria entered in 2007) that varied significantly in GDP/capita. But there was also a moderating factor: their (and EU-15) coal-dependent electric power-producing and energy-intensive industries were sectors covered by the ETS. This reduced lobbying and made the ESD politically easier than previous burden-sharing arrangements. Strong national and corporate interests linked to the ETS sectors were not directly affected by the upcoming ESD.

The sharing of the –10 per cent target for the non-ETS sectors was based on one criterion: GDP/capita. Member-states below the EU GDP/capita average were to have less ambitious targets, with a maximum limit on emissions: 20 per cent above 2005 levels. Member-states above the EU GDP/capita average would be expected to make greater efforts, up to a maximum of 20 per cent below 2005 levels. The individual targets could be achieved by a combination of unilateral and EU-wide policies and measures, such as the regulation on CO_2 emissions from new cars.

The ESD proposal not only shared the effort among the member-states: it also contained several mechanisms for promoting the 10 per cent reduction target. First, a linear reduction path was foreseen, as in the ETS, based on annual reporting for each member-state. Second, there was flexibility in meeting the target: member-states could borrow some of their allowed emissions from the following year, or bank emission reduction in excess of their targets for the following year. In addition, some CDM credits could be used for national goal attainment. Third, the Community could launch infringement procedures against non-complying member-states. This might allow the Commission, in extreme cases, to request the Court of Justice to impose financial penalties.

As for the RED proposal, the drafters had already put aside a mandatory trade-based certificate system compatible with the internal energy market (see Chapter 4). The RED proposal contained mandatory but differentiated national targets that aggregated to 20 per cent increase in renewable energy consumption by 2020. The targets were based on a flat rate of 5.5 per cent, with the remaining gap allocated according to GDP/capita. The European Council had already decided on a 10 per cent binding share of biofuels for all member-states. Other elements of the proposed RED included mandatory national renewable action plans on how to achieve the national targets, an optional cooperation mechanism whereby countries could share the burden by achieving their targets jointly, administrative and regulatory reforms, and criteria whereby only biofuels that met sustainability criteria would count towards the 10 per cent biofuels target.

The CCS Directive was the world's first international legal framework for safe capture and storage of carbon. Acknowledging that fossil fuels would continue to play an important role in the energy mix, the Commission had based its proposal on the Communication on sustainable power generation from fossil fuels where it had become clear that the EU needed both a legal framework and mechanisms to provide incentives for CCS. The proposal addressed the former point. It contained a regulatory framework on risk management, licensing and liability. No specific mechanisms for funding CCS were included beyond the expected carbon price incentive, as operators would not have to purchase stored emissions.[1] The CCS proposal responded particularly to the needs of the fossil-fuel industries and the CEECs. Electricity production in Poland, for example, was based almost entirely on indigenous coal.

The efforts required would have to be considered fair, as some member-states were more able than others to finance the necessary investments in low-carbon solutions. The main criterion for calculating differentiated national targets in ESD and RED was therefore GDP/capita. Overall costs to EU economies were estimated at just under 0.5 per cent of GDP by 2020. In principle, no member-state was expected to undertake investments that diverged too sharply from this average. The EU climate and energy package was based on a thorough impact assessment of how the ETS, ESD and RED proposals would work together to level the costs and enable the costs to be shared (Commission 2008b). Significant member-state differences in costs by 2020 were levelled out by side-payments to compensate poorer member-states, so as to make the package politically acceptable. In essence, this was to be done in various ways:

- different national targets in the non-ETS sectors (ESD) based on GDP/capita;
- different national targets for the share of EU energy consumption to be achieved by renewables based on a combination of GDP/capita and flat-rate increase in the share of renewable energy;
- auctioning revenues (ETS) to compensate lower-income member-states.

Flexibility mechanisms were included in the proposal for new state-aid guidelines.[2] These would guarantee that member-states could grant generous operating aid to the production of renewable energy and energy-efficient co-generation, as well as providing legal certainty as to the EU ETS, and opening the possibility that state aid for CCS might be considered.

Commission President Barroso could be said to have travelled far on climate policy since 2005, when in January 2008 he co-presented the climate and energy package with Environment Commissioner Stavros Dimas, Energy Commissioner Andris Piebalgs and Competition Commissioner Neelie Kroes, who had proved in practice that they were able to put together an ambitious package.

5.2 NEGOTIATING THE PACKAGE

The package was well received by the Parliament and the Council. To a large extent, this can be understood in light of the integrative nature of the package, promoting cost-effectiveness, fairness in burden-sharing, and mutually reinforcing climate and energy goals. In addition, the 20–20–20 targets supported by the Parliament and adopted by the European Council in 2007 required an ambitious package of measures if words were to be translated into action. On the other hand, there was no guarantee that the Commission's specific legislative proposals would survive further negotiations.

In fact, the main structure of the package remained intact throughout the 2008 negotiations. Member-states unanimously adopted the package and the Parliament endorsed it – and also the reduction targets for ETS and non-ETS sectors and differing national targets in the ESD and RED, as well as the CCS proposal. The four legislative proposals were complex, with a tight schedule for their adoption. The EU needed an ambitious package to show its 'leadership by example' before the international climate negotiations in Copenhagen and the June 2009 Parliament elections might result in a less supportive Parliament. Deliberations commenced in the first half of 2008 with informal trialogue negotiations among high-level representatives from the Commission, the Parliament and the Council. Various procedural challenges were sorted out, such as how to move forward on sustainability criteria for biofuels applicable to the RED and FQD.

The Core Package

Negotiations intensified in autumn 2008. French Presidency leadership proved extremely important in forging compromise by the end of that year. One institutional tool the French government used was the introduction of a more demanding decision-making procedure, replacing qualified majority voting in the Council of Ministers with unanimity in the European Council (Skjærseth and Wettestad 2010). In principle, each member-state plus the Parliament would have a veto position. Because the elements of the package were mutually reinforcing, decision-makers were pressured to negotiate and adopt *all* legislative proposals simultaneously.

There were various areas of disagreement in the negotiations on reforming the ETS. The new CEECs, led by Poland, demanded more economic 'solidarity' than already included in the Commission's proposal, and free allocation of allowances for the electric power sector. The energy-intensive industries demanded more free allowances and access to CDM credits than proposed, to reduce the risk of their re-locating outside Europe (Skjærseth and Wettestad 2010). These demands were voiced with increasing intensity in autumn 2008, fuelled by the unfolding global economic crisis and rising concerns about the costs of the package. One challenge was that these demands were conflicting: the more allowances distributed for free, the fewer revenues would be available for a solidarity fund for the CEECs. Moreover, credits from CDM would reduce incentives for low-carbon investments within the EU. The allocation issue was also linked to the international climate negotiations. If the Copenhagen Summit in December 2009 did not succeed in providing an ambitious international agreement, the EU would have to protect its energy-intensive industries by free allowances or a border tax to ensure a level playing field.

Poland coordinated its position with the Visegrad Group (V-4), that also included the Czech Republic, Hungary and Slovakia.[3] Prior to the EU leaders' meeting in October 2008, concessions to Poland were prepared in the context of the unfolding financial crisis; Italian Prime Minister Silvio Berlusconi called for a 'pause for reflection' (ENDS 2008c). The Commission had proposed 10 per cent of the revenues to the 'solidarity fund', whereas Poland was calling for 30 per cent. In November, the Baltic states joined the V-4 alliance and teamed up at a coordination meeting in Warsaw (ENDS 2008d). The coalition of CEECs threatened to veto the whole package if the energy situation in these countries were not taken better into account (Ancygier 2013, p. 126). To prevent this, Commission representatives travelled to Warsaw to 'sell' the climate and energy package by emphasizing synergies and new low-carbon opportunities, like CCS. In December, the Polish Prime Minister held a summit with French President Sarkozy. The meeting did not resolve the 'solidarity fund' issue, but a compromise was reached on granting power stations a longer derogation period from auctioning, from 2016 to 2020 (ENDS 2008e). The Polish position was that free allowances to protect energy-intensive industries exposed to carbon leakage should apply to the electricity sector as well.

On 12 December 2008, a final compromise was reached on the EU ETS. First, Poland and other CEECs won concessions postponing the phase-in of auctioning for power plants. In 2013, certain stations would get up to 70 per cent free allowances, decreasing towards zero by 2020.[4] These derogations from the main auctioning principle ended up as Article 10c of the revised EU Emissions Trading Directive. However, in exchange for free allocation, the CEECs would have to develop national plans to make their electricity sectors more efficient and also diversify their energy mix through investments worth at least the same as the value of the free allowances.

Second, regarding solidarity, 88 per cent of allowances to be auctioned each year would be distributed to the member-states. Ten per cent would be distributed to poorer member-states, as proposed by the Commission. Two per cent would go as a bonus to member-states whose GHG emissions in 2005 were at least 20 per cent below their 1990 emissions: the CEECs. Still, those concessions did not meet the CEEC demands; Poland was particularly dissatisfied with the outcome. Agreement could not be reached on the mandatory use of revenues for low-carbon investments, as some member-states saw this as an exclusive national competence.

Third, the European Council agreed on the less stringent requirement of a 70 per cent auctioning rate by 2020 in industrial sectors not exposed to the risk of carbon leakage, starting from 20 per cent in 2013. Energy-intensive sectors exposed to the risk of carbon leakage would receive 100 per cent free allowances if these sectors conformed to product-based benchmarks set for the best available technology. Energy-intensive industries would also gain greater access to external CDM/JI credits than originally proposed by the Commission, as companies would have roughly similar access in the third trading period as in the second trading period (Skjærseth and Wettestad 2010).

Germany in particular defended the energy-intensive industries and successfully demanded more free allowances than proposed by the Commission. The Parliament initially opposed this (see Chapter 6), but conceded after gaining acceptance for its demand to use 300 million allowances from the ETS New Entrants' Reserve – emissions rights for new entrants under the ETS – to co-finance up to 12 CCS demonstration projects and other renewable technologies, linking the proposed CCS legislation to the ETS.

The negotiations on the CCS proposal thus introduced significant changes in how to incentivize CCS. The Commission's proposal included only the ETS carbon price, as operators would not have to purchase allowances for stored emissions. The Parliament rapporteur for the CCS dossier – Chris Davies – managed to shift the focus towards making CCS commercially viable (Chiavari 2010). Davies cooperated closely with Avril Doyle – rapporteur for ETS revision, who had presented an amendment to set aside 500 million carbon allowances. Davies also proposed CO_2 emission limits on power stations, to force the use of CCS on coal power (Chiavari 2010). Several member-states opposed the Parliament proposals, particularly coal-dependent new CEECs that foresaw high CCS costs. After lengthy and complex negotiations, 300 million was the compromise solution. Negotiations thus brought in issues that were not part of the CCS proposal and solutions that were made part of negotiating another proposal: the ETS revision. The flexibility provided by simultaneous negotiation of different policies made possible this trade-off in concessions.

In order to reach a political compromise, the negotiations on the revised ETS and CCS policies were also linked to new state-aid rules under the competence of DG Competition. The Commission issued a statement setting out the main principles it intended to apply in new state-aid guidelines on compensation for indirect costs due to the EU ETS (costs in the form of higher electricity prices) in order to avoid undue distortion of competition (see Commission 2012a). And, at the European Council meeting on 11 and 12 December 2008, the Commission issued a statement that would allow the member-states to use revenues from the auctioning of allowances to support the construction of new, highly efficient power plants and CCS-ready power plants (Commission 2009).

With regard to the ESD, the final outcome followed the main lines of the proposal. Importantly, the individual targets for each member-state based on 2005 emission levels remained intact. Nevertheless, the CEEC group demanded changes in the baseline and structure of the proposal that were not compatible with the overall design of the package. These demands were counted in the concessions granted the CEECs in the ETS negotiations on increased revenues from auctioning and free allowances for the power sector, linking negotiations in the two policy proposals (Skjærseth 2013, 2014). In addition, the final decision introduced more flexibility in reaching the individual targets (Lacasta et al. 2010). Member-states can apply for inclusion of non-ETS activities or substances in the ETS sectors. This provision creates a direct link between the revised ETS and flexibility in the ESD. Further, member-states short on Annual Emissions Allocations (AEAs) for the non-ETS sectors can purchase from countries with a surplus.

Throughout the negotiations on the RED proposal, both the differentiated national targets and the 10 per cent minimum target for renewable energy in the transport sector were upheld. Most disagreements had been sorted out in the drafting period in 2007. Some member-states still feared that the national targets would be too ambitious, and held that the Renewable Energy Directive should allow for maximum flexibility. The flexibility mechanism proposed by the Commission, whereby member-states could coordinate target achievement, remained a voluntary option.

Several technical details as to how the national targets were calculated were changed to assuage member-state concerns, including how to calculate the renewables share (Howes 2010). This had consequences for countries with large fossil-fuel-based energy-producing industries, such as Norway, the Netherlands and Poland. Energy consumption within energy-producing industries was mostly excluded from the basis for calculating. This in turn made it easier to attain a higher renewables share (Bøeng 2010).

Some of the most contentious issues were related to biofuels and sustainability criteria. These negotiations became extremely complex, as sustainability criteria should not diminish the competitiveness of European industry or lead to trade barriers – the latter was important, as import of biofuels would be necessary to achieve the target. Moreover, since sustainability criteria would be part of both RED and FQD, consistent criteria were needed (see below).

According to the Commission proposal, only biofuels that met sustainability criteria would count towards the biofuels target. This caused heated debate among the member-states. Three blocs emerged: countries favouring strict criteria, like Denmark, Germany and to some extent Sweden; countries like Poland, Austria, Spain and France, which preferred lax criteria; and those that were indifferent, as were many of the new member-states (Müngersdorff 2009, p. 28).

Members of the European Parliament proposed many amendments, with strong initial disagreements within the ITRE Committee that were sorted out in hasty compromises because of the tight deadline. The ITRE Committee finally agreed to strengthen the sustainability criteria, demanding higher GHG-savings thresholds for biofuels (Müngersdorff 2009). While reluctant member-states dismissed these higher thresholds, the European Parliament managed to get some of its amendments through into the final Renewable Energy Directive: a substantial review clause instructing the Commission to assess various issues further, including the minimum GHG emissions-saving threshold; and to make the review clause applicable to the overall 20 per cent renewable energy target and not only to the biofuels target (Müngersdorff 2009).

Negotiating Decoupled Transport Policies

The Fuel Quality Directive was proposed in 2007 – before the climate and energy package proposal. Negotiations on the FQD became related to the 2008 negotiations because demands were made that it, like the RED, should include binding sustainability criteria for biofuels. This was partly settled in February 2008, when EU governments created a working group to establish common criteria. The European Parliament rapporteurs on both renewables and fuel quality supported this move; an alliance of member-states and MEPs was thus created to pressure the Commission to include binding sustainability criteria also in the FQD (Christensen and Gulbrandsen 2012).

The final text of this Directive was agreed in November 2008 and presented as part of the climate and energy package. Included was a requirement for fuel suppliers to reduce life-cycle GHG emissions from road fuels by 6 per cent from 2010 to 2020 (Article 7a), which introduced a reduction target for GHG emissions from fuels. For biofuels to count against the GHG emissions reduction targets they must meet certain sustainability criteria set out in the FQD, corresponding to those in the RED. However, the method for calculating life-cycle emissions from fuels other than biofuels remained an unresolved issue postponed for further comitology negotiations. Part of the reduction could come from the purchase of CDM credits.[5]

Negotiations on a new Car Emissions Regulation proposed by the Commission in December 2007 took place alongside the negotiation of the 'core' package. The Commission proposed 2012 as the target date for reducing CO_2 emissions from cars to 120 g CO_2/km. Improvements in motor technology would have to reduce average emissions to no more than 130 g CO_2/km, while the remaining 10 g would be left to softer 'complementary measures' (Commission 2007). During the negotiations, the regulation was aligned with the package by the inclusion of a longer-term emissions limit of 95 g CO_2/km by 2020. This marked the end of the voluntary approach.

Germany and France dominated these negotiations in the Council. A major point of disagreement between them was whether the curbs to apply should be a car's weight (favoured by German carmakers) or the size of its carbon footprint (favoured by French carmakers). Since their interests were so divergent, it was widely assumed that any deal reached between these two countries could serve as the basis for an EU-wide agreement.

Meeting for the annual bilateral summit in June 2008, German Chancellor Merkel and French President Sarkozy sorted out some of their differences. Both backed the 120 g/km by 2012 target, saying that the targets should be based on vehicle weight. In addition, both called for a 'substantial' phasing-in of the target and extra incentives for eco-innovations, as well as greater flexibility on penalties (ENDS 2008b). This bilateral deal was a major concession on the part of France, which produces smaller, less-polluting cars compared with Germany. Achieving an agreement on the car regulation was a priority for France, due to assume the EU Presidency in July, as it wanted to bolster the EU's leadership role in addressing climate change (ten Brink 2010, p. 197). This can help to explain why France was ready to yield to German demands (particularly the weight parameter) – the Presidency wanted to 'close the deal' before the end of 2008 (Gulbrandsen and Christensen 2014, p. 517).

Reactions to the German–French car deal were varied. German carmakers and the ACEA welcomed the deal. Environmental NGOs, however, were enraged, declaring that the deal ruined the chances of an ambitious car emissions regulation. On 4 November 2008, negotiations began between the Parliament and the Council under the French Presidency, with the Commission acting as mediator. The Council and the European Parliament agreed to gradually limit CO_2 emissions from 130 g/km for 65 per cent of new cars in 2012 to 100 per cent by 2015. The target of average emissions at 95 g CO_2/km by 2020 was also agreed. As with the FQD, several details were postponed for further deliberations in comitology.

5.3 THE WAY FORWARD

The climate and energy package aimed at strengthening the credibility of EU climate-policy leadership internationally. 'If we want a global agreement it's absolutely indispensable that Europe leads the way,' declared Barroso when the package was proposed (ENDS 2008a). The prospects of an ambitious international agreement were directly related to the EU's climate target and policies. If a new and sufficiently ambitious deal could be concluded, the EU would adjust its GHG reduction target upwards to ensure an overall 30 per cent reduction in ETS and non-ETS sectors. This would in turn allow for more import of external credits from the international flexible mechanisms. In the absence of international progress, the 20 per cent by 2020 target would remain, and energy-intensive industries would be protected from carbon leakage by free allowances based on benchmarks.

The conditions set by the EU for moving up to a 30 per cent GHG reduction target were not met at the Copenhagen Summit. The 'Copenhagen Accord' did not set global reduction targets; it did not add up to what scientific advice deemed necessary for remaining within the 2°C objective, and it was not legally binding. The Accord fell short of the EU's expectations – but it provided a basis for further work. The 2011 Durban Platform for Enhanced Action aimed at finalizing a new 2020 climate agreement with legal force by 2015. The Doha 2012 outcome included a new commitment period for the EU under the Kyoto Protocol, which corresponded to EU targets and policies adopted for 2020 (Skjærseth 2013).

The failure to adopt more ambitious international climate commitments meant that remaining internal adjustments to the climate and energy package could be sorted out in comitology, on the basis of the '20 per cent by 2020' goal. Adjustments made to the ETS and ESD served to further increase flexibility and reduce the pressure on the member-states and energy-intensive industry. The Commission reached a final Decision that extended the number of ETS product benchmarks from 42 to 52, plus three more for fuel and heat consumption and certain process emissions (Commission 2011a). This extension reflected heavy pressure from energy-intensive industries. Guidelines for state aid to cover indirect costs from the ETS were presented by DG Competition in late 2011 and were finally adopted by the Commission in 2012. These covered also rules for aid involved in the transitional free allowances for the modernization of electricity generation in the CEECs (Commission 2012a). In October 2012, the national limits in the non-ETS sectors covered by the ESD were approved. The Decision set AEAs in tonnes for each member-state for each year from 2013 to 2020. The Decision also stated that the maximum quantity of emissions for each member-state should be adjusted in accordance with changes in coverage of the Emissions Trading Directive. The NER 300 funding mechanism was implemented by a Decision specifying criteria and measures for financing commercial demonstration projects (Commission 2010a).

Car and fuel-quality legislation adopted in 2008 left contentious issues for comitology. In the end, the car legislation was watered down. The Act was finally adopted by the Parliament on 25 January 2014 and the

Council on 10 March 2014. As of this writing, the FQD methodology has not yet been decided. A highly problematic issue was whether (and when) to add a default value for oil (tar) sands distinct from that for conventionally produced oil. Moreover, the sustainability criteria for biofuels had to be specified, spurring debate on how to calculate the impact of *indirect* land-use change (ILUC).[6] On 6 October 2014, the Commission finally adopted a new proposal to implement existing obligations in the FQD, Article 7a in particular.[7] The Commission proposal has yet to be approved by the Council and the European Parliament. In retrospect, we can see how this underlines the political importance of decoupling these policies from the 'core' package: legislation on fuel quality and car emissions involves powerful corporate and member-state interests.

The financial crises unfolding from 2008 lowered emissions in the ETS and non-ETS sectors, and reduced implementation costs as well as the incentives for long-term low-carbon investments. The Commission in 2010 presented an analysis of options for unilaterally moving beyond 20 per cent GHG reduction through a strengthening of the climate and energy package, or new additional policies (Commission 2010b). The Communication concluded that a 30 per cent cut by 2020 was technically feasible and economically affordable, and discussed the means and consequences of moving to this target in light of the new circumstances. This meant that the climate and energy package would need to be renegotiated – risking the fragile consensus reached in 2008. The 30 per cent unilateral target was discussed internally throughout 2010. It was pushed by Climate Commissioner Connie Hedegaard and DG Climate Action,[8] and supported by the UK, Ireland and Denmark. The French and German governments were split, with their environmental ministers supporting a more ambitious carbon target, while ministers of the economy were more sceptical. In particular, Poland and Italy had been long-standing opponents of tougher emissions cuts. Vigorous opposition came also from the European Alliance of Energy Intensive Industries, which argued that the higher target would be detrimental to the competitiveness of its members. In a resolution adopted prior to the international climate negotiations in December 2010 in Cancún, the Parliament plenary called on the EU to adopt a unilateral 30 per cent target (European Parliament 2010).

However, lack of full support from member-states and industry stalled such plans. The Commission then used another European Council agreement from 2009 as a foundation for stepping up policies: the goal of reducing GHG emissions by between 80 and 95 per cent by 2050 compared with 1990 levels (European Council 2009). In March 2011, the

Commission issued a roadmap prepared by DG Climate Action for moving towards a competitive low-carbon economy by 2050 (Commission 2011b). It showed that GHG emissions would have to be reduced by 25 per cent in 2020, 40 per cent in 2030 and 60 per cent by 2040 (all in terms of 1990 levels) to reach 80 per cent reduction by 2050. The roadmap encouraged decision-makers to strengthen the EU ETS and raise carbon prices: by setting aside allowances to provide scarcity in the market and by revising the 1.74 per cent linear-reduction factor. Poland, supported by energy-intensive industries across the EU, blocked the roadmap and subsequent efforts at finding a compromise (ENDS 2011; European Council 2011, p. 3).

Within the Parliament, responses to the low-carbon roadmap were divided. The Environment Committee voted overwhelmingly in favour of maintaining the 30 per cent CO_2 reduction goal for 2020. The Committee argued for a resolution whereby, under the current 20 per cent target, the ETS would have a very limited role in driving emissions reductions and the deployment of low-emission technologies (European Parliament 2011, p. 4). At the plenary session of the Parliament in June, however, centre-right MEPs supported the roadmap target of a 25 per cent cut. Since then, the Parliament has voted in favour of the roadmap and the strengthening of the EU ETS (European Parliament 2012).

In December 2011, the Commission followed up with an Energy Roadmap 2050 prepared by DG Energy[9] (Commission 2011c). Here the key message was that greater energy efficiency and more use of renewables to achieve 80–95 per cent reduction by 2050 would cost about the same as would continued heavy reliance on nuclear power and fossil fuels. This roadmap proposed increasing the share of renewables to around 30 per cent by 2030. It further stressed that decarbonization of the EU should not develop in isolation, but take into account international developments, to minimize carbon leakage and adverse effects on competitiveness (Commission 2011c).

In June 2012, Poland vetoed also a compromise proposal on the Energy Roadmap, arguing that EU efforts should be matched internationally and that references to 'decarbonization' should be deleted (European Council 2012). Poland did not see any market opportunities in renewables (see Chapter 7). Several delegations also wished to delete a reference to a greater than 30 per cent share of the energy consumption mix for renewable energy in 2030; Germany, Belgium, Ireland and Sweden, however, wanted to keep this reference (ENDS 2012; European Council 2012).

Failure to agree on a more ambitious 2020 climate target and stepwise targets towards 2050 pitted EU climate- and energy-policy instruments against each other and raised tensions between DG Climate Action and DG Energy. It became increasingly clear that climate and energy policies could be conflicting. In addition to the effects of the economic crisis, greater shares of renewables and more energy efficiency in the ETS sectors would reduce demands for EU emissions allowances (Commission 2012b, 2012c). This was acknowledged in the new Energy Efficiency Directive (EED) issued in October 2012, which included explicit references to the need to strengthen also the EU ETS in order to make it more effective.

Something had to be done to rescue the EU's climate-policy flagship. The use of international credits had added to the surplus of allowances, which had accumulated to 955 million allowances by early 2012. This, combined with low demand, sent the carbon price plunging, from nearly €30 in spring 2008 to just above €5 in spring 2012, with a significant reduction in the second half of 2011. The surplus was expected to build up in 2012 and 2013, to reach over 2 billion allowances in phase 3.[10] In July 2012, the Commission initiated legal changes to postpone or backload auctioning of 900 million allowances from the beginning to the end of the 2013–2020 period (Commission 2012d). The proposal proved extremely controversial but was finally adopted in December 2013. Some believed that postponing of auctioning would not work. Backloading was expected to stabilize the carbon price but not solve the surplus problem, as it would not remove the allowances permanently from the system (Commission 2012e). Others saw backloading as an unjustified intervention in the carbon market. Most energy-intensive industries resisted any measures to fix the ETS – despite the Commission's previous efforts at sweetening the pill for these industries by proposing a list of sectors that could be given state aid to assist in meeting the indirect costs of the ETS.

5.4 CONCLUSIONS

The outcome of the decision-making phase proves to be mainly in line with our expectations to issue-linkage 'success' expectations, as outlined in Chapter 2. Linkages between issues and policies offered joint gains for decision-makers that could veto new climate and energy policies that differed from the status quo. The package of policies proposed by the Commission in January 2008 provided ample room for mutual concessions among pivotal actors in the negotiations. The Parliament got a new funding mechanism for CCS based on the ETS revenues for accepting more free allocations to energy-intensive industry, as demanded by

Germany in particular. Energy-intensive industry gained more access to external credits. Poland and other CEECs were granted more subsidies from emissions allowance auctioning revenues, as well as several other derogations from accepting the ESD. Still, the agreement reached in 2008 proved to be a complex and politically fragile matter.

The need for 'give-and-take' in the negotiations to secure more ambitious policies was reduced by the way in which the new policies were linked. First, the package was designed to combine different concerns for climate change and energy security, to promote synergies with air pollution and 'green' economic growth, and not least to provide side-payments to compensate less wealthy member-states. The proposed package was based on common impact assessment of the various package components where the aim was sharing the costs according to GDP/capita. Second, issues and policies that might obstruct the basis for agreement were decoupled from the core package and negotiated independently. These included energy efficiency, car emissions and fuel quality. The two latter proved extremely politically contentious, and have undergone various renegotiations. Third, complexity was reduced by postponing several matters for later negotiations. The most contentious issue here concerned the benchmarking criteria for energy-intensive industries.

Lack of progress on new international climate commitments and the emerging financial crisis changed the situation shortly after the package was adopted, and led to several new policy initiatives. These initiatives failed, because renegotiation of the package could threaten the fragile consensus that had been achieved. Thus, we see that one drawback of making complex negotiating outcomes based on policy packages is that it may prove difficult to adapt the agreement to new circumstances. Here, the only measure that eventually succeeded was a regulation on postponing the auctioning of 900 million allowances to stabilize the carbon price at a low level. New strategies for 2050 garnered broad support but were all vetoed by Poland.

The package was proposed by the Commission at member-state request, as expected by the LI approach. But the outcome was not independently shaped by negotiations among the member-states, and it only partly reflected the constellation of interests among the various member-states. The initiatives to NER 300 and a more stringent review of sustainability criteria for biofuels were taken by the Parliament. While the European Council had decided on the targets and the main principles, the ETS reform was largely crafted by the Commission, which had initiated the ETS in the first place. An EU-wide cap and auctioning had figured high on the Commission's list ever since 2000.

On the other hand, the outcome was not independently shaped by the Commission proposal and by negotiations between the Parliament and the member-states, as could be expected from the MLG approach. The RED and the ESD and the negotiations on car emissions and fuel quality were mainly the products of intergovernmental negotiations. The RED proposal clearly anticipated the positions of the member-states after the Commission's failed attempt to get acceptance in 2007 for a mandatory and common trade-based certificate system.

Therefore, we conclude that both these perspectives on EU policymaking can help to explain different components of the package. As EU energy policymaking has traditionally been more intergovernmental in nature than has climate policy, the combined energy and climate package came to reflect a blend of LI and MLG.

Factors external to the EU also affected the adoption of the package. The economic crisis, unfolding from the autumn of 2008, strengthened arguments for reducing the costs of the package through higher imports of external credits. The crisis also determined the EU policy initiatives taken after the package was adopted, most notably the backloading decision. Further, the international climate negotiations placed a commitment and time-pressure on the 2008 negotiations which helped to promote success in the negotiations. Conversely, the failure to agree on a new international climate treaty made new policy initiatives more difficult. And finally, the Kyoto Protocol's commitment periods and the direct links to compliance with the ETS, FQD and the ESD through the purchase of external credits served to lower abatement costs for countries and for industries.

History is full of targets and policies that are never realized. In the next chapters, we examine the implementation of the EU's climate- and energy-policy package in Germany, Poland, the Netherlands and Norway.

NOTES

1. However, operators were made responsible for stored CO_2 for a certain period, where in case of a leak they would have to purchase allowances.
2. The new state-aid rules entered into force on 1 April 2008.
3. The V-4, established in 1991, is a formal body for cooperation on various issue-areas, with a rotating presidency.
4. Eligible installations would be those that were poorly integrated in the European grid or that individually provided more than 30 per cent of national electricity in countries with relatively low GDP.
5. The legislation applies to all petrol and diesel used in road transport, as well as gasoil used in non-road-mobile machinery. The 6 per cent was presented as part of a bigger 10 per cent reduction target, consisting of a 6 per cent reduction in GHG intensity of fuels by 2020, with intermediate indicative targets of 2 per cent by 2014 and 4 per cent by 2017; an

additional 2 per cent reduction subject to developments in new technologies such as CCS; and a further 2 per cent reduction to come from the purchase of CDM credits.
6. While emissions from *directly* converting land to agricultural use for producing biofuels (direct land-use change) must be reported under the revised FQD adopted in 2009, modelling is necessary to estimate the degree of ILUC.
7. The new proposal establishes a method for calculating the carbon intensity of various fuel types: petrol, diesel, Liquefied Petroleum Gas (LPG) and Compressed Natural Gas (CNG). Each of these types will be assigned a default value based on emissions over its entire life-cycle. Suppliers must use these values when reporting the carbon intensity of their fuel supply to member-states to ensure a 6 per cent reduction in their transport emissions in 2020.
8. In February 2010, departments relating to climate change were split off from the environment DG to form the DG Climate Action.
9. In February 2010 Energy was split from DG TREN to form the new DG Energy.
10. These include forward-selling to generate funds for the NER 300 programme, early auctioning to meet power-sector hedging demand, and the sales of left-over allowances in phase 2 NER.

REFERENCES

Ancygier, A. (2013), *Misfit of Interests instead of the 'Goodness of Fit'? Implementation of European Directives 2001/77/EC and 2009/28/EC in Poland*, Hamburg: Verlag Dr. Kovac GmbH.

Bøeng, A.C. (2010), Konsekvenser for Norge av EUs fornybardirektiv. *Økonomiske analyser* 4: 48–58, accessed 7 January 2016 at http://ssb.no/a/publikasjoner/pdf/oa_201004/boeng.pdf.

Brink, P. ten (2010), 'Mitigating CO_2 emissions from cars in the EU (Regulation (EC) no 443/2009', in S. Oberthür and M. Pallemaerts (eds), *The New Climate Policies of the European Union: Internal Legislation and Climate Diplomacy*, Brussels: Brussels University Press, pp. 179–209.

Chiavari, J. (2010), 'The legal framework for Carbon Capture and Storage in the EU', in S. Oberthür and M. Pallemaerts (eds), *The New Climate Policies of the European Union*, Brussels: VUB Press, pp. 151–77.

Christensen, A.R. and L.H. Gulbrandsen (2012), *EU Policies on Car Emissions and Fuel Quality: Reducing the Climate Impact from Road Transport*, FNI Report 14/2012, Lysaker: Fridtjof Nansen Institute.

Commission (2007), *Results of the Review of the Community Strategy to Reduce CO_2 Emissions from Passenger Cars and Light-Commercial Vehicles*, COM (2007) 19 final, 7 February, Brussels: European Commission.

Commission (2008a), *Europe's Climate Change Opportunity*, COM (2008) 30 final, 23 January, Brussels: European Commission.

Commission (2008b), *Impact Assessment, Document Accompanying the Package of Implementation Measures for the EU's Objectives on Climate Change and Renewable Energy for 2020*, SEC (2008) 85/3, 23 January, Brussels: European Commission.

Commission (2009), *Commission Decision Responding to the Request of ClientEarth to the European Commission for an Internal Review of the Commission Statement and Article 10 Paragraph 3 of the EU ETS Directive (2008/0013(COD)) Adopted by the European Parliament on 17th December 2008*,

Brussels, 27 April 2009, accessed 13 October 2015 at http://ec.europa.eu/environment/aarhus/pdf/client_earth_reply.pdf
Commission (2010a) 'Commission Decision of 3 November 2010 laying down criteria and measures for the financing of commercial demonstration projects that aim at the environmentally safe capture and geological storage of CO2 as well as demonstration projects of innovative renewable energy technologies under the scheme for greenhouse gas emission allowance trading within the Community established by Directive 2003/87/EC of the European Parliament and of the Council', *Official Journal of the European Union*, L 290/39, 6 November.
Commission (2010b), *Analysis of Options to Move Beyond 20 Per Cent Greenhouse Gas Emission Reductions and Assessing the Risk of Carbon Leakage*, COM (2010) 265 final, 26 May, Brussels: European Commission.
Commission (2011a), 'Commission Decision of 27 April 2011 determining transitional Union-wide rules for harmonised free allocation of emission allowances pursuant to Article 10a of Directive 2003/87/EC of the European Parliament and of the Council', *Official Journal of the European Union*, L 130/1, 17 May.
Commission (2011b), *A Roadmap for Moving to a Competitive Low Carbon Economy in 2050*, COM (2011) 112 final, 8 March, Brussels: European Commission.
Commission (2011c), *Energy Roadmap 2050*, COM (2011) 885 final, 15 December, Brussels: European Commission.
Commission (2012a), 'Guidelines on certain State aid measures in the context of the greenhouse gas emission allowance trading scheme post-2012', *Official Journal of the European Union*, C 158/4, 5 June.
Commission (2012b), *Commission Staff Working Document: Proportionate Impact Assessment*, draft, accessed 13 October 2015 at http://ec.europa.eu/clima/policies/ets/cap/auctioning/docs/swd_2012_xx2_en.pdf.
Commission (2012c), *The State of the European Carbon Market in 2012*, COM (2012) 652 final, 14 November, Brussels: European Commission.
Commission (2012d), *Commission Prepares for Change of the Timing for Auctions of Emission Allowances*, 25 July, accessed 19 May 2015 at http://ec.europa.eu/clima/news/articles/news_2012072501_en.htm.
Commission (2012e) *Commission Submits Draft Amendment to Back-load 900 Million Allowances to the Years 2019–2020*, accessed 13 October 2015 at http://ec.europa.eu/clima/news/articles/news_2012111203_en.htm.
ENDS (2008a), 'Businesses' contrasting reactions to climate plans', 24 January.
ENDS (2008b), 'France and Germany unite to delay car CO_2 curbs', 10 June.
ENDS (2008c), 'Climate legislation "not a luxury EU can forgo"', 15 October.
ENDS (2008d), 'Alliance against parts of EU climate plans widens', 6 November.
ENDS (2008e), 'Governments glum at Poznan climate talks', 8 December.
ENDS (2011), 'Too much or too little? Reactions to EU carbon and energy plans', 9 March.
ENDS (2012), 'Poland vetoes energy roadmap to 2050', 15 June.
European Council (2009), *Presidency Conclusions*, 29/30 October, Brussels, 1 December.

European Council (2011), *Presidency Conclusions*, Brussels: General Secretariat, 22 June.

European Council (2012), *Draft Council Conclusions*, Brussels: Permanent Representatives Committee, 8 June.

European Parliament (2010), Climate: EU should move to 30 per cent emissions reduction target, say MEPs, Brussels: Press release.

European Parliament (2011), *On the Analysis of Options to Move beyond 20 Per Cent Greenhouse Gas Emission Reductions and Assessing the Risk of Carbon Leakage*, PR\858607EN.doc, Brussels: Parliament.

European Parliament (2012), *Parliament Calls for Low-Carbon Economy by 2050*, PR\40876, Brussels: Parliament.

Gulbrandsen, L.H. and A.R. Christensen (2014), 'EU legislation to reduce carbon emissions from cars: intergovernmental or supranational policy-making?', *Review of Policy Research*, 31 (6), 503–28.

Howes, T. (2010), 'The EU's Renewable Energy Directive', in S. Oberthür and M. Pallermarts (eds), *The New Climate Policies of the European Union*, Brussels: VUB Press, pp. 117–50.

Lacasta, N., S. Oberthür, E. Santos and P. Barata (2010), 'From sharing the burden to sharing the effort', in S. Oberthür and M. Pallermarts (eds), *The New Climate Policies of the European Union*, Brussels: VUB Press, pp. 93–116.

Müngersdorff, M. (2009), *Evaluating EU Decision-Making Processes – A Case Study on the 'Renewable Energy in Transport Target' and the 'Sustainability Criteria for Biofuel Production' in the Context of the EU's Climate and Energy Package*, Working Paper, Stockholm: Stockholm Environment Institute.

Skjærseth, J.B. (2013), *Unpacking the EU Climate and Energy Package: Causes, Content, Consequences*, FNI Report 2/2013, Lysaker: Fridtjof Nansen Institute.

Skjærseth, J.B. (2014), 'Linking EU climate and energy policies: causes, content, consequences', *International Environmental Agreements: Politics, Law and Economics,* 2014. Published online 2 October 2014, 17 p. DOI: 10.1007/s10784-014-9262-5.

Skjærseth, J.B. and J. Wettestad (2010), 'Fixing the EU Emissions Trading System? Understanding the post-2012 changes', *Global Environmental Politics*, 10 (4), 101–23.

6. Implementation in Germany

Per Ove Eikeland

6.1 INTRODUCTION

Germany represents a particularly interesting case for the study of EU decarbonization efforts. Germany is the largest economy in the EU; its carbon-intensive energy supply[1] makes the country by far the largest emitter of GHGs: 21 per cent of total EU emissions in 2012 (European Environment Agency 2015). Historically, Germany has championed climate-policy development at national and EU levels (Jänicke 2010). This, together with restructuring after German reunification, brought GHG emissions into a falling long-term trend (Eichhammer et al. 2001; European Environment Agency 2011). Germany continued its champion role in 2007 when it adopted national GHG emissions and renewable energy targets more ambitious than those collectively negotiated and decided at the EU level. Against this background, we would expect good EU policy implementation performance in Germany – an expectation that this chapter can only partly confirm.

6.2 GERMAN CLIMATE AND ENERGY POLICIES UP TO 2007

Three key objectives drove German energy policies in the baseline period: affordable energy; environmental protection and sustainability; and security of supply (Duffield and Westphal 2011). Affordable energy was essential for maintaining the industrial structure of manufacturing industries in value chains with energy-intensive basic materials industries that provide much of total national value-added and employment. Its exceptional priority was reflected in the German ministry structure: responsibilities for both industry and energy policies were initially gathered in one lead ministry, the Bundesministerium für Wirtschaft und

Energie – BMWi). Coal (anthracite or hard coal, and lignite) remained an important affordable energy source for ensuring national industrial development.

Attention to security of supply has increased in tandem with higher energy-import dependencies, but has also experienced variation with events in the international energy market (Duffield and Westphal 2011). The ratio of net imports/total primary energy supply increased from 48 per cent in 1990 to around 60 per cent in 2007, when nearly all the oil, 80 per cent of natural gas and one-third of the coal needed for meeting demand was imported (International Energy Agency (IEA) 1990, 2007).

Environmental concerns evolved as the premise for energy policy in the 1970s (notably, air pollution from coal plants). Concerns about nuclear radiation spurred a strong German anti-nuclear movement in the 1980s, gaining force after the 1986 Chernobyl disaster. The police clamped down on local civil disobedience protesters, causing deep distrust in the major utilities and their allies in the federal government. The political parties were split, and a new Green Party (established in 1983) became the first of its kind to win seats in a national parliament in Europe. A red/green coalition government of the Social Democratic Party (SPD) and the Green Party, formed after the 1998 elections, negotiated a 'nuclear consensus' with the major utilities that essentially entailed phasing out nuclear power by 2022. German political parties could note greater consensus for an ambitious climate policy, reflecting demands from an environmentally conscious population. Ambitious national GHG emissions reduction targets were adopted in the late 1980s (a 25 per cent cut by 2005, with 1987 as the base year). In 1997, Germany's Kyoto target was set at a 21 per cent cut by 2008–2012 from 1990; in 2002 the red/green government agreed to a 40 per cent cut by 2020 if the EU as a whole committed to 30 per cent (Karlseng 2006). Climate- and energy-policy discussions became strongly interconnected, reflecting differing party-political views on the continued use of nuclear power as a low-carbon source of energy.

Under the German 'ordo-capitalist' economic development model, a core function of the state was to support industrial development. Industry associations and trade unions were given priority access to policymaking, so that leading scholars have classified Germany as a country with strong neo-corporatist elements (Katzenstein 1987; Vitols 1997). Policies were often settled in negotiations between the government and these groups. Negotiated agreements became an important element of the German energy- and climate-policy mix as well, as did 'command-and-control' regulations like demands for 'best available technology'. These were combined with generous state support for investments: favourable long-term investment loans provided by semi-private investment banks (like the Kreditanstalt für Wiederaufbau – KfW) and major federal and state government investment programmes (energy-efficiency measures, CCS and renewable energy) (Ziegler 1989). Such 'economic carrot' instruments created a parallel demand for new domestic technology, in turn stimulating growth in the country's important manufacturing industry.

'Economic sticks' (beyond transport fuel taxes) had a moderate place until the Ecological Tax Reform in 1999. In 2005, Germany introduced trading in emissions allowances as part of its implementation of the EU 2003 Emissions Trading Directive. Manufacturing and coal-power generation sectors were given major rebates from environmental taxes and refund opportunities that sometimes reduced the effective tax rate to almost nil (Heine et al. 2012) and large surpluses of allowances under German implementation of the EU ETS (Eikeland 2013).

One of the most generous 'economic carrot' instruments adopted in Germany was the feed-in tariff system for operation support for renewable electricity. The policy idea emerged locally: it was proposed by backbenchers in the Bundestag in the early 1990s and enacted against the will of traditionally influential industry groups – showing that neo-corporatist policymaking was not all-dominant (Lauber and Mez 2004). Policy-strengthening after 2000 was facilitated after responsibility for renewable energy policy was shifted from the BMWi to the Ministry for the Environment (Bundesministerium für Umwelt – BMU) in 2002 (Lauber and Mez 2004; Suck 2005). Investments by private households, farmers, businesses and municipal energy companies – not the major power companies – led to rapid growth.[2] Another generous policy was tax exemption for biofuels and support for German farmers for biofuel feedstock production, leading to high growth rates in this area.

Conflicts between the BMWi and the BMU became a recurrent feature of German renewable energy and climate policymaking. Added to this came centre–periphery conflicts that put their mark on the country's energy and climate policies: under the federal political system, 16 individual German state (*Länder*) governments have co-legislative powers in federal lawmaking (Bulmer et al. 2003; Lauber and Mez 2004).

The European Union played a growing role as an arena for Germany to upload ambitious EU GHG emissions reductions and renewable energy targets: to deal with the climate problem, to avoid competitive disadvantages for own industry if regulated more strictly, and to extend the market for German 'green' technologies. Germany promoted renewable electricity and fuels directives at the EU level (Lauber and Mez 2004), but played the role of sceptic when the European Commission (henceforth:

Commission) drafted its flagship energy and climate policies. Germany contributed to curtailing effective internal energy-market policies that would dismantle national anti-competitive practices; to stopping the adoption of EU-harmonized allocation rules under the ETS; and stopping an internal energy-market compatible system for supporting renewable electricity at the EU level.

6.3 NEGOTIATING THE PACKAGE: 2007–2008

A coalition government of the parties CDU/CSU[3] and SPD held power in Germany when the EU proposed and negotiated its energy- and climate-policy package in 2007–2008. This coalition government supported the triad 20–20–20 by 2020 targets, with primary preference for a 30 per cent target for GHG emission cuts, and also favoured making this and the renewable energy target binding. The German government provided significant inputs to drafting and negotiating the package. Important background factors here were the ambitious national climate and energy targets for 2020 that the government agreed on in August 2007 (the 2007 Meseberg Declaration). These included a 40 per cent cut in national GHG emissions and a series of sector-based renewable energy targets: a 25–30 per cent share of electricity consumption (12 per cent in 2006); a 14 per cent share of heat consumption (6 per cent in 2006); and a 17 per cent share of transport fuels consumption (6 per cent in 2006). Other targets were a doubling of national energy productivity by 2020 compared with 1990, with the subtarget of doubling the share of electricity generated in CHP plants to 25 per cent by 2020 (Federal Environment Ministry 2007). The Declaration committed the government to work for an EU Car Regulation with the average 120 g/km-standard by 2012 proposed by the EU (10 grammes to be achieved by measures outside the vehicle); a reformed EU-level ETS; and to provide EU-level incentives for CCS, based on the intention of taking the lead in EU technology development with two or three demonstration plants constructed on German soil. Like the EU package, this 29-point policy programme aimed at balancing concerns for security of supply, economic efficiency and environmental protection.

German Environment Minister Sigmar Gabriel endorsed the Commission-adopted proposal for a revised EU ETS Directive, on condition that energy-intensive industries exposed to international competition would continue to receive CO_2 allowances free of charge. Germany opposed postponing the identification of sectors at risk of leakage and setting benchmarks for these, and kick-started assessment work together with the British and Dutch governments. A preliminary assessment report by the Commission was not accepted by Germany, which wanted more generous coverage of sectors. When the final benchmark scheme was adopted by comitology in April 2011, the German government won through on more demands than it had expected. Germany was also a strong supporter of the provision that gave member-states the right to compensation to competition-exposed sectors for indirect costs of the EU ETS. New state-aid guidelines to this end were adopted by the Commission in May 2012, and Germany later gained acceptance for its national compensation scheme. Germany further won through with its demand for the right to opt out of the proposed centralized trading platform and set up a national trading platform instead. However, Germany failed to gain sufficient support for binding earmarking of revenues from the auctioning of allowances to climate projects, and the European Council only agreed to recommend voluntary earmarking of 50 per cent of the revenues. Germany also supported specific earmarking of revenues to CCS demonstration projects – the NER 300 mechanism (Boasson and Wettestad 2013).

As for the Commission-adopted proposal for an ESD, the German government accepted the suggested 14 per cent national target. Germany had no prior differentiated targets for ETS and non-ETS sectors in its national policy, but the total national GHG emissions reduction target combined with sector targets for renewables appeared more ambitious than the EU set target for the non-ETS sectors.

As noted, the German government was generally positive to creating an EU-level environmental/safety framework for CCS projects, and to piloting such projects funded by the EU within Germany. This position reflected the high share of coal in German electricity production (49 per cent in 2007), with coal investments on the increase due to the decision to phase out nuclear power, lignite resources as an affordable and abundant domestic energy source, and German companies already pioneering pilot CCS projects on coal power-plants (Eikeland 2013). The government thus regarded CCS as an essential long-term solution for cutting GHG emissions in Germany. A contested point in the negotiations concerned liability rules – the time that project operators would be fully responsible for monitoring and preventing leakage from storage points before handing these responsibilities over to national authorities. Germany wanted a longer liability period for operators (50 years) than the adopted 20 years. This was, however, still in line with German preferences, as long as the final Directive granted member-states discretion to set a longer liability period.

The German government was strongly in favour of binding renewable energy targets differentiated for the member-states, and accepted the 18 per cent total target (share of total energy consumption) and specific 10 per cent target for renewable transport fuels, although opting for a higher target of 12.5 per cent. National sector targets set in 2007 were more ambitious. Germany lobbied heavily (and successfully) for full opt-out from a proposed EU-level renewable electricity certificate trade, instead supporting the series of flexibility mechanisms available to member-states for assisting national target achievement.[4] Germany supported sustainability standards for biofuels as part of the Renewable Energy Directive and the Fuel Quality Directive, as well as life-cycle-based emissions cut for fuel suppliers, as proposed by the latter (ENDS 2008).

The German government also supported the Car Regulation, including binding EU targets for manufacturers, the integrated approach (biofuels should be allowed, used as a measure for target achievement) and targets differentiated for classes of cars (explained in Chapter 5). The Commission-adopted proposal was generally in line with German government preferences, including the burden-sharing mechanism that was tilted towards benefiting manufacturers of the large cars typically produced in Germany. Germany also uploaded amendments that extended phase-in dates, granted more leeway in the penalties applied to producers, and added that specific eco-innovations could be accorded extra weight in target achievement. The German government accepted that a 95 g CO_2/km for 2020 would also become binding, on condition that the Commission first conducted a review of the legislation in 2013 and presented new details for meeting this target. In further comitology negotiations, Germany won through with further demands on flexibility and postponement in timetables for reaching this target (see Chapter 5.3).

All in all, Germany preferred, and EU legislation also adopted, ambitious and binding GHG and renewable energy targets; an EU ETS with generous free allocation to energy-intensive industry and compensation for higher electricity prices; no EU-harmonized tradable renewable electricity certificate system; a moderate car regulation adapted to the German car manufacturing industry; and emissions-cut targets for fuel suppliers/sustainability standards for biofuels. With the exception of the EU GHG reduction target of 20 per cent, which was less ambitious than that preferred by Germany (30 per cent), the country basically got what it wanted in the negotiations. And that should lead us to expect good implementation of all parts of the package.

6.4 IMPLEMENTING THE PACKAGE

German implementation (transposition and application) took place under shifting coalition governments in the period 2009 to 2014, all headed by Chancellor Angela Merkel of the CDU. The grand coalition CDU/CSU/SPD government that negotiated the package began delivering its legislative proposals to the Bundestag under the parallel 2007 national Integrated Energy and Climate Programme (IECP) and drafting transposition measures once the EU package was settled in late 2008. After the Bundestag elections in 2009, the Merkel II conservative–liberal coalition government (CDU/CSU and Free Democratic Party (FDP)) continued policy implementation. In 2010 this government agreed on new long-term 2050 renewable energy and energy-savings targets ('the Energy Concept') combined with postponement of the nuclear power phase-out date. The latter was revoked after the Fukushima nuclear plant disaster in 2011, but the long-term *Energiewende* targets for 2050 were retained: 60 per cent share of renewables in total gross energy consumption; 80 per cent share of renewables in electricity supply; 50 per cent reduction in primary energy consumption; 40 per cent reduction in final energy consumption; and 25 per cent reduction in final electricity consumption. The annual rate of energy-retrofitting buildings was doubled, to 2 per cent. The subsequent Merkel III grand coalition government (CDU/CSU and SPD), formed after the 2013 elections, upheld the 40 per cent national GHG emissions reduction target for 2020 and announced that it would work to get this target adopted by the EU as a whole for 2030. The government parties agreed to continue the *Energiewende*, now formulated with legally binding corridors for expansion of renewable electricity: 40 per cent to 45 per cent by 2025 (replacing 35 per cent by 2020 and 50 per cent by 2030 under the Renewable Electricity Sources Act); and 55 per cent to 60 per cent by 2035 (replacing the previous goals of 50 per cent by 2030 and 65 per cent by 2040) (Federal Ministry for the Environment, Nature Conservation and Nuclear Safety 2011).

EU ETS

The Merkel II government undertook timely and correct legal transposition of the EU ETS Directive in the form of a reformed Greenhouse Gas Emission Trading Act in July 2011 (Eikeland 2014). It also decided to transfer 100 per cent of the revenues from auctioning allowances to a new special energy and climate fund. Timely transposition of the Directive notwithstanding, the German government overstepped the 30

September 2011 deadline for reporting its national implementation measures (NIMs) – the plan for preliminary allocation of allowances to the installations covered by the system in the period 2013–2020. The government submitted its NIMs report only after the Commission had launched infringement procedures in late April 2012. The Commission found minor nonconformities with EU benchmarks rules, and prevailed in its demands on Germany to accept a cut in allowances for some companies (Eikeland 2014).

When the Commission in 2012 proposed postponing the auctioning of 900 million allowances from 2013 to 2015 as a measure aimed at upward correction of the collapsed allowance price (backloading), the German government failed to agree on a position. This was alleged to have created uncertainty at the EU level, also delaying the decision process here (ENDS 2012). By contrast, the new grand coalition government in 2013 agreed to support the backloading proposal.

During its first year of operation in 2013, the revised EU ETS Directive did not bring about any noticeable emission cuts from Germany's ETS sectors. Emissions from energy installations increased by 2 per cent despite a major deficit of allowances, even after free allocation was almost completely removed. Low carbon prices and low international coal prices caused a surge in coal-power output. Emissions from installations owned by energy-intensive industries decreased by 1 per cent despite the vast surplus of allowances transferred from the second trading period (Deutsche Emissionshandelsstelle 2014).

Against the backdrop of falling allowance prices that failed to incentivize emission cuts, in 2011 the government signed new voluntary agreements with the industry and energy industry associations, instructing companies to implement certified energy management systems and prove annual energy-intensity reductions in exchange for continued energy tax rebates (Eikeland 2014). In response to studies showing that Germany would miss its national target of 40 per cent GHG cuts by 2020, the government in December 2014 announced a new carbon-policy action plan that would include a cap on emissions from the electricity sector for the period 2016–2020 to be added with a national emissions-trading system for this sector, subject to a legality check under German constitutional rights and EU law. That proposal has met stark opposition from the electricity industry (German Energy Blog 2015b).

Effort-Sharing Decision

The ESD did not require formal transposition in German law. However, national policies and measures would have to be implemented if

Germany were not to miss its mandated GHG emissions reductions of 14 per cent by 2020 for the non-ETS sectors (transport and buildings sectors as main emitters). Germany has proposed and adopted several policies for these sectors that variously represented also implementation efforts for the 10 per cent renewable transport-fuels target under the Renewable Energy Directive, the Fuel Quality Directive and the Car Regulation. Assessment indicated that, as of 2014, Germany was unlikely to achieve the ESD target by 2020 unless additional national measures could be implemented (European Environment Agency 2014). This reflects implementation problems for a series of policies targeting the building stock and transport sectors in Germany.

Policies to stimulate more fuel-efficient and carbon-leaner cars included amendments to the Motor Vehicle Tax Act in 2009, after rounds of re-drafting in the government and mediation between the Bundestag and the Federal Council, which wanted compensation to the *Länder* for loss of tax revenues. The amendments made small fuel-efficient cars cheaper, but more powerful vehicles not much more expensive (*Spiegel* Online, 2009). Electric vehicles (EVs) were given a five-year tax exemption, and a National Development Plan for Electric Mobility set up the goal of one million EVs on German roads by 2020. A €500 million funding programme for R&D and market introduction targeting German car manufacturers was established for the period 2009–2011. The Merkel II government in 2010 re-launched the plan, extending funding programmes to €2 billion and adding the goal of six million EVs on German roads by 2030. This still represented less than 10 per cent of the car stock registered in Germany by 2014.

Car manufacturers, the environmental movement and *Länder* governments continued lobbying for extended tax exemptions, with a limited 10-year exemption period proposed as an amendment to the Motor Vehicle Tax Act in July 2012 for cars purchased between 18 May 2011 (when amendments were first promised) and 31 December 2015. Following further criticism that policy incentives were insufficient for consumers to purchase EVs, plug-in hybrid and fuel-cell vehicles, a September 2014 bill proposed expanding the number of measures, to allow municipalities to reserve parking spaces for re-charging stations, provide free parking, distribute special access and transit passes in areas exposed/sensitive to air or noise pollution, and open bus lanes for such vehicles (EurActiv 2014).

All in all, implementation problems for German transport policies made sector emissions continue growing (IEA undated; EurActiv 2015). Concerning policies specifically addressing the uptake of CO_2-lean cars, Germany was in late 2014 still far from its target of one million EVs on the road by 2020 – there were only about 19,000 EVs and 107,000 hybrid cars registered (German Energy Blog 2015a). This being said, German car manufacturers had clearly reduced the average CO_2 emissions of new registered cars: by 2013, this average was 138 g, down from 157 g in 2009. German manufacturers offered about 880 models on the German market that emitted at most 130 g CO_2/km; and 528 models under 120 g CO_2 (Verband der Automobilindustrie 2014). German industry was thus well on track as regards complying with the EU Car Regulation target set for 2015. Meeting the 95 g CO_2/km target set for 2024 has remained a major challenge. According to the industry, this target can be met only by selling a considerable number of cars powered by units other than the traditional combustion engine (Verband der Automobilindustrie 2014). Also, difficulties in meeting the 10 per cent renewables transport-fuel target under the Renewable Energy Directive contributed to the lack of overall progress in reducing emissions from the transport sector, as discussed below.

Policies to stimulate emissions cuts from heating the building stock – the other main non-ETS emitting sector – included ambitious 2020 targets for energy savings for new buildings, amended building standards, and extended preferential loans and programmes that were enacted in 2009. The following government set higher targets for energy savings – including renovation rates for existing buildings and climate-neutral building standards for new building by 2020 and long-term targets for 2050. It also promised extended funding programmes, to be financed by ETS allowance auctioning revenues. However, these experienced cuts instead, because of the economic crisis and the collapse of EU ETS allowance prices. Moreover, political implementation problems emerged, as when the federal government in 2011 proposed tax deductibility for energy-efficient renovation of existing buildings. The Federal Council opposed the bill, fearing loss of tax revenues for the *Länder* governments. Mediation with the Bundestag on compensation did not succeed, and the bill was finally rejected in December 2012, with the launch of a less comprehensive new KfW preferential loan programme instead.

Renewable Energy Directive

The RED set the binding target for Germany to source 18 per cent of its total energy consumption and 10 per cent of transport fuel consumption from renewable energy by 2020. As the targets set energy/fuels consumption as the denominator, achievement would be facilitated also by energy/fuels savings. German transposition of the RED basically meant the continuation of existing policies, including the amendments planned and new subsector targets set in the 2007 ICEP.

Policies aimed at achieving the national renewable electricity goal (25–30 per cent by 2020) included 2009 amendments to the Renewable Energy Sources Act which provided for extended feed-in tariffs for various technologies as the principal policy instrument. An exception was solar photovoltaic (PV) power, for which faster tariff degression rates were implemented, to reflect rapid cost reductions for this technology, which was experiencing high growth rates but also rapidly increasing costs for consumers who paid a surcharge to finance the system. A new ordinance gave large surcharge rebates to German energy-intensive companies, to shield them from the costs of the feed-in tariff system.

Policies aimed at achieving the national renewable heat target (14 per cent by 2020) included a new Renewable Energy Heating Act of 2009 (instructing homeowners to cover heating needs by renewable energy or better insulation/heat from CHP) and investment support through the long-existing Market Incentive Programme. This programme was scaled up from €237 million (2008) to €426 (2009) (Federal Ministry for the Environment, Nature Conservation and Nuclear Safety 2013).

Policies on renewable transport fuels (to achieve the national target of 17 per cent by 2020) included as their main instrument mandatory blending of biofuels with conventional fuels (2006 Biofuels Quota Act), which had replaced earlier full tax exemption for biofuels in place since 2002. New sustainability standards for biofuels (required by both the RED and Fuel Quality Directive) were implemented swiftly, made applicable from November 2009. To ensure compliance, unions and organizations of the German agrarian and biofuel industry created a certification system in 2010.

The subsequent Merkel II government submitted its National Renewable Energy Action Plan (NREAP) to the Commission on time, reporting that reforms of the feed-in tariff system would be implemented. The NREAP did not specify any use of flexible cooperation mechanisms with other countries. The Commission accepted the German transposition of the RED but noted that the German NREAP accounted well for national measures only in renewable electricity and heat, not renewable fuels in the transport sector.

Reform of the generous feed-in tariff system proved difficult: investments in renewable electricity boomed, but so did costs to consumers and fears that a growing volume of intermittent power could cause problems in balancing the electricity system. Changes to the solar PV tariffs were proposed in 2010, 2011 and 2012, in the form of tariff cuts, setting up expansion corridors and demands for direct marketing for solar PV producers. Some changes were accepted only after Federal Council modifications that did not stop the investment boom. More comprehensive reforms for all renewable electricity technologies were first outlined in 2010, to replace the feed-in tariff system with a market-adapted system (direct marketing instead of automatic guaranteed feed-in, market premiums and tendering instead of set tariffs). A first-step reform proposal was included in the post-Fukushima policy package of May 2011, whereby investors could choose an optional market premium instead of traditional feed-in tariffs, accepted by the Bundestag and Federal Council after modifications. More general reforms became postponed to after the 2013 elections because of internal government disagreements. The subsequent Merkel III coalition government agreed on a reform bill that would require direct marketing for all plants by 2017, auctioning from 2018 to determine the level of support, a cut in average support across all technologies from 17 ct/kWh to 12 ct/kWh for new installations by 2015, and target corridors for all technologies with financial support adjusted if investments were higher or lower than these corridors.

Reforms were supported by the Commission, which saw the feed-in tariff system as obstructing the internal electricity market. The Commission directly interfered with the German reform process by initiating investigation of whether the reduced feed-in surcharge rates for selected German electricity-intensive manufacturing enterprises and rail operators represented a breach with EU state-aid rules, and by the parallel preparation of new state-aid guidelines that would exclude the use of feed-in tariffs. The German government pursued its national reform in parallel with lobbying the new EU framework together with other governments and industries. These efforts bore fruit: extended timelines and derogations were entered in the state-aid guidelines, and there was continued acceptance for exempting industries from paying full renewable energy surcharges. The reform bill was adopted only after new concessions were accorded to the *Länder*. In response, the head of the Commission's energy service, Günter Oettinger, stated that Germany would need far more fundamental reforms to Europeanize its renewable energy policy (*PV Magazine* 2014).

Germany's approach to assimilating the Renewable Energy Directive with existing national renewable electricity policies resulted in rapid growth for renewable electricity: its share of total electricity consumption rose from 14 per cent in 2007 to 27 per cent in 2014. In the boom years 2010 to 2012, more than 7 GW of solar PV alone was added each year. However, the 18 per cent target for Germany under the RED is measured as share in total energy consumption. Here Germany's figures are less impressive, up from 7.5 per cent in 2007, 10.4 per cent in 2010 and 12.4

per cent in 2013, still a way to go to reach the 18 per cent target. This reflects slow uptake of renewables in heating and transport sectors. The share of renewables in German heat consumption levelled off in the period 2010–2013 at around 10 per cent, up from 7.5 per cent in 2007 (Federal Ministry for the Environment, Nature Conservation and Nuclear Safety 2013). The German federal government sought to correct for this by amending the Renewable Heating Act in 2011, increasing the share of heating in buildings to be covered by renewables (up from 15 per cent to 25 per cent). Challenges remained, however. The amendments also stipulated a hardship clause, whereby municipalities in financial difficulties could be exempted from the obligation. And federal support of investments under the Market Incentive Programme declined after the collapse in EU ETS allowances prices which funded the programme (Federal Ministry for the Environment, Nature Conservation and Nuclear Safety 2013).

The share of renewables in total German transport fuel consumption fell sharply, from 7.4 per cent in 2007 to 5.7 per cent in 2012 (Federal Ministry for the Environment, Nature Conservation and Nuclear Safety 2013). This reflected the fact that tax exemptions for biofuels were replaced by a new law mandating new biofuel blends that were difficult for the market to accept and adopt. One example is implementation of the new fuel standard E10 (10 per cent ethanol blended in petrol), delayed because of poor information from fuel suppliers as to which cars could run on the new standard, and thus lack of consumer trust in the new blend. Green campaigns questioning the climate soundness of the new fuel added to customer scepticism.

CCS Directive

Under the 2007 IECP, the German government pledged speedy adoption of a legal framework to ensure a stable basis for investments and operation of CCS facilities in Germany, and stating its intention of changing the Federal Regional Planning Act so as to secure significant locations for CO_2 storage across the country. The most promising were in the *Länder* of Schleswig-Holstein, Niedersachsen and Brandenburg.

The Merkel I government finished drafting a transposition bill tailored to full-scale CCS but failed to present it to the Bundestag before the 2009 elections because of local opposition. The ensuing CDU/CSU/FDP coalition government retained the commitment to CCS but re-wrote the bill with strict limits as to the volumes of CO_2 that could be stored, adapted for demonstration rather than full-scale plants, and with various new provisions that included possible financial compensation for affected communities, and an opt-out clause allowing the *Länder* to designate/ban areas for CCS pilot projects, on the basis of specific criteria. This concession to *Länder* governments opposed to the bill was contested by the government of Brandenburg, a major lignite-extraction *Land* and also home to CO_2 storage-testing projects. The Brandenburg government feared that a ban on CO_2 storage in the geographically more suited North German *Länder* would put a stop to the continued use of coal in the future decarbonized energy system (*Die Welt* 2011).

Political stalemate postponed government adoption of the bill to April 2011. It was approved by the Bundestag in July but rejected by the Federal Council in September, at which point the transposition deadline had already been missed. In late June 2012, mediation between the Bundestag and the Federal Council ended in a compromise that further scaled down annual limits on storage volumes and extended the post-closure obligation period for storage operators. This, and the opt-out clause for the *Länder*, meant that none of the initially planned demonstration plants would come on stream. Already in December 2011, Vattenfall AG, operator of the most mature CCS plans, announced the cancellation of its NER 300-sponsored demonstration project which would have included storage in the *Land* of Brandenburg. This put a halt to federal government plans to adopt the CCS Directive so as to enable EU-supported demonstration plants to be located on German soil.

Conclusions

Germany experienced delays in implementation of the CCS Directive and the NIMs report under the EU ETS Directive. The German CCS transposition resulted in the cancelling of industrial development initiatives that were underway. The Renewable Energy Directive was transposed correctly and on time, but entailed cumbersome reform processes. Despite the considerable challenges in the areas of renewable heating and renewable fuels, Germany was well on its way to achieving the 2020 total renewable energy target, thanks to massive investments in renewable electricity. In 2014, reforms of renewable electricity policies slowed the growth in investments; and Germany was not entirely on track as regards meeting the ESD target because of challenges in getting policies adopted for the transport and building-stock sectors. Low carbon prices have contributed to recent increases in emissions also from the ETS sectors. Thus, Germany is unlikely to achieve its self-imposed 40 per cent GHG emissions reduction target unless it can succeed in implementing additional planned policies – which, given the short timeframe, appears challenging. With strong support in the population and important parts of its industry base, Germany has formulated plans for a national and EU transformation to a low-carbon energy system by 2050, and has started implementing policies towards this goal.

6.5 EXPLAINING IMPLEMENTATION CHALLENGES

Fit and Adaptation Pressure

The goodness of fit between EU requirements and the German situation – assessed in terms of the 'distance' between pre-existing policies, the energy-economic situation, negotiating positions and the final negotiated EU outcome – has been generally high, with some minor variation across specific EU climate and energy policies.

Concerning the EU ETS, Germany had been reluctant to accept the system in the early years of the new millennium, fearing that high carbon prices could harm the competitiveness of important national industries. Implementation in Phases I and II involved NAPs with abundant allowances given to these industries (Eikeland 2013). By 2008, Germany had adapted the ETS to its national policy mix, even becoming a frontrunner as regards auctioning of allowances in the electric power industry. Germany thus supported the continuation of the EU ETS in a revised form if carbon leakage provisions were included (free allowances and state aid to compensate for higher ETS-induced electricity prices). This was accepted, with no major mismatch between what the German government wanted and the final outcome. Germany's delays in submitting its NIMs represented a minor mismatch that stemmed from using a hardship clause as a basis for allocating additional allowances to a small number of installations, justified by the principle of proportionality enshrined in the German Constitution but not paralleled in the EU ETS Directive (Eikeland 2014).

Goodness of fit was also the situation as regards Germany and the CCS Directive. CCS demonstration projects had already been initiated, and the government was a clear advocate of rules that could create predictability for investors and decarbonize coal power – the latter accounted for around 50 per cent of total electricity supply. Government plans involved securing at least two EU-funded demonstration plants in Germany and exploiting the opportunities to make the country a technology leader. Poor fit cannot explain the considerable implementation problems experienced with this Directive.

Also the Renewable Energy Directive showed a good fit with the German situation. More ambitious renewable energy sector targets had already been adopted at the national level, and a harmonized EU-level trade-based support scheme was removed from the RED after massive lobbying by Germany and a group of allied countries. Also sustainability standards for biofuels (part of the Fuel Quality Directive as well) resonated well with policies already under development. However, despite the fit with the adopted package, Germany's implementation policies did not fit with parallel EU internal energy market policies, and the Commission exerted reform pressure. This poor fit may help to explain the initiation of reforms, but not wholly the outcome of the reform.

The ESD target for Germany of 14 per cent cannot be said to represent a poor fit that created adaptation pressure, if judged against the ambitious national 40 per cent total GHG emissions-cut target for 2020. Moreover, despite no prior national specification of ETS and non-ETS sector targets, Germany accepted the new EU target; and parallel national plans in 2007 aimed at strengthening non-ETS climate policies. It also accepted additional policies in the EU package aimed at facilitating emissions cuts from road transport (the Car Regulation), after major modifications more in line with Germany's interests as the major car-manufacturing country in the EU. Thus, poor fit is ill-suited for explaining German implementation problems regarding policies for GHG emissions cuts from the non-ETS sectors.

Our data do not indicate any major concessions given by Germany in the negotiations that resulted in greater adaptation pressure for parts of the package. The package as a whole fitted well with the energy-economic situation of the country and represented what it had sought. Germany had long strived to upload its ambitious national GHG emissions targets to the EU and global levels. With energy-import dependencies steadily increasing, Germany wanted the policy package to stimulate indigenous energy production. As home to the EU's largest manufacturing industry that was exposed to international competition, Germany wanted – and got – provisions that could protect this industry, in line with its earlier policies. Finally, Germany had become the most important market in the EU for the emerging 'green' industry in Europe, and that meant new industrial activity, employment and export value in line with the ecological modernization approach formulated for Germany's national industrial policy (Jänicke 2010). Ambitious EU targets for renewable energy were thus seen not only as a measure for GHG mitigation and security of supply: they were also important for expanding the 'home' market for German energy-technology industries. All in all, then, poor fit between the EU package and the situation in Germany can explain very little of Germany's implementation challenges.

Domestic Politics

Governmental actors

Germany experienced changes in the composition of its federal governments involved in EU policy implementation. Three succeeding coalition governments (all holding a majority in the Bundestag) retained the CDU/CSU as majority parties with Angela Merkel continuing as Chancellor, while the minority parties in these governments shifted. The SPD was part of the Merkel I and Merkel III governments; the FDP joined the Merkel II government. All governments maintained the ambitious 40 per cent national GHG reduction target and commitment to a large share of renewables for a long-term low-carbon transition of the national energy system, reflecting relatively high party-political consensus. However, the governments differed somewhat in their views on the exact content of the low-carbon transition (the mixture of nuclear and renewable power), timetables for renewable energy expansion and nuclear power phase-out, support regime for renewables, and diagnosis of the consequences of ambitious German and EU energy and climate policies for German industry and competitiveness. When the FDP held the important BMWi ministerial post under the Merkel II government, the competitiveness of traditional German industries was given higher priority as an energy-policy objective, in line with party priorities. The FDP consistently favoured less rapid growth in renewable energy and a market-based support system, postponement of the nuclear phase-out and low ETS allowance prices to counter the costs of a low-carbon transition for Germany's traditional industries. By contrast, with the CDU and SPD parties there was more internal disagreement on policy reforms. With the FDP as minority party, the government could not agree on a position on the EU ETS backloading proposal, which delayed implementation at the EU level. The shift of minority partner to the SPD in 2013 cleared the way for agreement, and the reform was implemented. On the other hand, CDU/CSU ministers disagreed on the swift fundamental reforms of renewable energy policies proposed by the FDP, and these were postponed until after the elections in 2013. The reforms carried out with the CDU/CSU and the SPD in the new government were comprehensive but not as far-ranging as those proposed by the FDP.

However, Germany is also a federation of 16 individual *Länder*, with co-legislative powers in federal policymaking for *Länder* governments vested with the Federal Council (Bundesrat). Since elections to the *Länder* parliaments are not synchronized with federal elections, the political composition of the Federal Council will often differ from that of the Bundestag, giving German opposition parties a venue to veto policies even if they are in the minority at the federal level. None of the three Merkel governments has held a majority in both the Bundestag and the Federal Council, with local parties and governments also occasionally diverging from their federal groupings.

The power of the Federal Council differs, depending on whether federal government bills are *consent* or *objection* bills. A *consent* bill can be rejected if it is opposed by the Federal Council and if further mediation with the Bundestag does not result in a solution accepted by the Federal Council. However, most bills, including energy and climate bills, are *objection* bills. These do not require consent by the Federal Council, although objections and calls for mediation with the Bundestag may delay lawmaking. If such bills are objected to by an absolute majority or two-thirds majority of the Federal Council, then the Bundestag must match such a majority if it is to disregard the objection.

Transposition problems for the CCS Directive mark a clear case where consent was needed because it interfered with the exclusive rights of the *Länder* in area planning and where such consent was not given by the Federal Council. Strong objections were voiced by the governments of Schleswig-Holstein and Lower Saxony, even though they were headed by Angela Merkel's CDU Party, which favoured the Directive. These two *Länder* had the greatest potential for underground storage of carbon in Germany, and had been targeted for storage by several industrial initiatives. Underpinning these objections were widespread popular lack of trust in carbon storage as a safe option, and massive anti-CCS campaigns that portrayed CCS as a saviour for the German coal industry and thereby a potential obstacle to the renewable energy transformation, which had already advanced rapidly in these northern *Länder*. Their governments convinced also other colleagues in the Federal Council not to accept a law that would interfere with regional prerogatives in area planning. Thus, the final transposition law determined that the *Länder* would have the right to ban future underground storage of carbon. The government of Lower Saxony has since presented a bill on banning any permanent CO_2 storage in the *Land*, and Schleswig-Holstein has also voiced this as a possibility.

Reforming renewable energy policy is an illustrative example of the use of Federal Council powers even when formal consent was not needed. Throughout the implementation period 2009 to 2014, the Federal Council delayed reforms by invoking Mediation Committee negotiations. In one instance, a two-thirds majority of the Federal Council – including representatives of *Länder* of the same political colour as the federal government – objected to the reforms. Such a high vote gave the Federal Council real veto power, as the Bundestag could not muster a similar two-thirds vote to neutralize the objection. The parties of the Merkel II government had only a 53 per cent majority in the Bundestag; this increased to nearly 80 per cent under the Merkel III government, improving the capacity for renewable energy policy reform. Underlying causes for *Länder*–government opposition included strong popular backing for the German energy restructuring project and the fact that the new renewable energy industry had brought industrial growth and employment opportunities (Eikeland 2014). The Federal Council voting system also contributed to other implementation problems, as with the veto of a bill on tax incentives to promote energy-efficient renovation of old buildings, important for target achievement under the ESD.

Affected societal actors

Looking at cost and benefit consequences from implementing the EU package for various societal groups, we can note considerable costs for the major *electric power utilities* – such as no more free allowances under the ETS, loss of market share and lower revenues/asset values due to the RED. On top of this came decommissioning costs from the decision to phase out nuclear power. *The Economist* (2013) reported that net income for Germany's top utilities was down by a third from 2010 to 2013, with plunging share prices. Other consumer groups also experienced high costs, as was the case for households and companies that were not exempted from paying feed-in surcharges that escalated since they were designed to increase when market prices went down. Moreover, households would get no compensation for indirect costs of the EU ETS or gain exemptions from paying various energy and electricity taxes. On the other hand, some groups emerged as true winners, notably businesses specializing in manufacturing, sales and installation of green technology (renewable power technology in particular).

By contrast, Germany's major energy-intensive companies experienced few costs from the EU ETS. There was abundant free allocation of allowances, compensation for higher electricity prices due to the EU ETS, generous exemptions from paying the feed-in surcharge for expanding renewable electricity, falling wholesale market prices from 2011 because of this expansion, and additional exemptions from paying national CO_2 taxes.

Moreover, the German government included various compensation and side-payment schemes in its implementation package. All German electricity consumers (households and farmers, as well as companies) were provided with opportunities under the feed-in tariff system to invest in their own renewable energy capacity, thereby becoming exempt from paying the surcharge and with new revenue opportunities from selling surpluses back to the grid. The policy tradition of using economic carrots (investment subsidy and R&D programmes, preferential loans) to stimulate development and uptake of new technology (thus compensating for costs) was continued by also including 100 per cent of the revenues from auctioning allowances under the EU ETS, to the benefit of a great many sectors, although total revenues available through this mechanism declined with the plunge in allowance prices.

We find only partial support for the expectation that highly concentrated costs combined with distributed benefits would lead to opposition among high-cost groups, and thus implementation challenges. Some high-cost groups, like the big utilities, opposed the national mode of implementing the Renewable Energy Directive (the feed-in tariff system) but were supportive of making the EU ETS more effective (support for backloading) – the latter viewed as the lesser evil despite the potential extra costs for their fossil-fuel-based generation capacity. On the other hand, German households experienced high implementation costs in the form of higher electricity prices, but showed continued high acceptance of these costs (Agentur für Erneuerbaren Energien 2014). The Federation of German Industries (BDI), representing both cost-shielded and cost-exposed industries, teamed up with the utilities in demanding the scrapping of feed-in tariffs, making the EU ETS the main energy- and climate-policy instrument, but not so as to make it more effective through backloading. Strong support for the whole package of policies came from associations representing the renewable energy industry. Here we must not forget that German energy and climate policy included many compensatory opportunities for high-cost groups (investment funding and access to feed-in tariffs) that were utilized by many German households, farmers and companies. By contrast, the big utilities made very limited use of the opportunities under the feed-in tariff system to invest in renewable energy capacity, seeing this more as a threat that could undermine the value of their existing assets.

Summing up, Germany's implementation of the EU energy and climate package was facilitated by the policy mix that provided opportunities for cost compensation and contributed to maintaining high popular acceptance of ambitious energy and climate policies. Some sector interests, viewing the policy mix as a threat, did not make use of the compensatory mechanisms and opposed Germany's approach to implementing the package. In particular, one compensatory mechanism was not utilized – funding of CCS demonstration projects through the NER 300s, as such projects were cancelled after Germany's implementation of the CCS Directive removed this technology as an option for assisting fossil-fuel-based interest groups in decarbonizing their energy installations.

Policy style and access

The general impression is that a consensual policy style dominated German implementation of EU energy and climate policies, enabling both promoters and supporters of reform to express their views in federal policy consultations and informally, through venues at the federal and *Länder* levels. With the federal political system of multiple veto players, various sector interests had a real chance to influence policy. Wider access to German energy and climate policymaking had emerged from the early 1990s, replacing the dominant corporatist style of policymaking whereby a limited group of major industry associations enjoyed priority access to influence policies. As a result, initial renewable energy policies spurred a new industry that gradually gained economic strength, organizational clout and political leverage at the local and regional levels, fostering new industrial activity, cleaner energy production and employment. The industry was supported by a broad coalition of actors that were important constituencies for German political parties – environmental groups, farmers and local communities. Greater leverage at the federal level came when the Green Party entered government in the late 1990s. It assumed charge of political leadership on renewable energy policies by shifting responsibility from the BMWi to the BMU, headed by a Green Party minister. The flipside of the coin was a parallel loss of influence for traditional corporate groups, like the big German utilities. These now had to compete for political attention while simultaneously experiencing a crisis of trust stemming from low environmental credibility and recurrent negative media reports on nuclear power plant incidents and lawsuits against the government, on removal of villages to make way for brown-coal mines, and on reported windfall profits gained under the EU ETS. The EU climate and energy package served to consolidate the political leverage of the 'renewable energy coalition' by providing new opportunities for export-driven renewable industry expansion.

While German renewable energy policies were typical of consensual-style policymaking, there are also indications that some parts of the package were first subject to less broad consultation processes, and that the government later initiated broader consensus talks after seeing such massive opposition. This was the case for the CCS Directive. However, broader consensus talks were not enough to make implementation progress as the government wanted – showing that consensus-style policymaking is no guarantee of successful policy implementation if opposing interests have leverage vis-à-vis national veto players. On the whole then, we find no unambiguous connection between policy style and implementation performance.

Administrative organization

The BMU had the lead responsibility for transposing all but one part of the package, the CCS Directive, for which the BMWi was the lead responsible. Conflicts over energy and climate policies between the two ministries had long existed, as noted above. Conflicts continued also during the negotiations and implementation of the 2020 EU package, principally over the EU ETS Directive and renewable energy policies.

In 2002, the BMU took over responsibility for renewable energy policies from the BMWi, which was blamed for obstructing the development of policies agreed by the coalition parties of the government. The BMU speeded up policymaking (which strengthened the feed-in tariff system), leading to an investment boom in renewable electricity. The feed-in tariff system was further strengthened as part of implementing the Renewable Energy Directive. The BMWi emerged as a champion of cutting tariffs, ultimately scrapping the feed-in tariff system, which it deemed non-compatible with its primary task of adapting Germany to the EU internal electricity market and keeping a check on electricity prices for the sake of national industrial competitiveness. The BMU long remained a veto player against such reforms, as the system delivered low-carbon investments, all the more important when the plunge in allowance prices made it impossible for the parallel EU ETS to offer such incentives. Here it was the BMWi, as watchdog over German industrial competitiveness, that took on the veto role when reforms of the EU ETS were proposed in 2012 (backloading), delaying a German government position and implementation at the EU level. The Federal Chancellery and Chancellor Merkel restrained from intervening to enforce stronger horizontal coordination when ministries were headed by the junior coalition party, fearing this might doom the government.

The situation changed with the Merkel III grand coalition government after the 2013 elections. Responsibility for policymaking on renewable energy was re-transferred to the BMWi in order to coordinate this policy field more closely with national energy market policies. The new government emplaced ministers from the same political party (SPD) to head of the two ministries, presumably to avoid party-political conflicts from exacerbating conflicts already latent between the two ministries. And reform capacity apparently increased. Reforms of the German Renewable Energy Law instituted a more controlled expansion in renewable electricity, limited the escalating costs for German consumers, and gave crisis-stricken utilities more time to adapt to the *Energiewende*. In parallel, the government agreed on EU ETS reforms, with the new BMWi minister Gabriel more actively supporting policies to raise allowance prices.

Hence, we find some indications that horizontal administrative fragmentation contributed to delaying energy and climate policy implementation in Germany. However, vertical administrative fragmentation remained a greater challenge. *Länder*-level governments can undermine federal policies by means of their formal co-legislative powers through the Federal Council; moreover, lower-level governments can block implementation of federal policies already adopted. Of particular relevance are *Länder* and municipal government prerogatives in national spatial and local area planning and in authorizing construction of new energy facilities (Schmidt and Buehler 2007), as this arrangement may delay or hamper implementation of federal laws and national plans. Implementation of the CCS Directive illustrates this clearly. Here, *Länder*-level governments vetoed the establishment of full-scale CCS plants; moreover, some of them later proposed laws that may permanently ban any underground storage of CO_2 on their territories. Another relevant policy area is transmission grid extension projects, which the federal government considers necessary for executing the *Energiewende*. Planning and permissions in the field have been seriously delayed because of local opposition. To date, this has not actually curtailed investments in renewable electricity, but continued large-scale investments may be challenged if Germany cannot speed up grid investments as well.

6.6 CONSEQUENCES OF GERMAN POSITIONS FOR LONG-TERM EU ENERGY AND CLIMATE POLICY

Despite the various implementation challenges, Germany has maintained its commitments to the national long-term low-carbon energy transition (ambitious 2050 and interim targets) and to uploading similar ambitions at the EU level. Germany pushed for ambitious targets under the EU 2030 energy and climate framework (40 per cent GHG emissions reductions, 30 per cent renewable energy and 30 per cent energy efficiency), the latter two higher than those set by the European Council in October 2014. From late 2013 onwards, Germany also took a leading role in pressing for EU ETS reforms and for making the proposed new Market Stability Reserve applicable from 2017 instead of 2021. Essentially, Germany was opting for a climate and energy package similar to the 2020 package.

The German position may be viewed in light of experiences with implementing the 2020 package. The government's continued push for renewable energy targets in the 2030 package reflects synergies experienced in Germany: contributing to GHG emissions reductions, containing growing energy-import dependencies, and stimulating new industrial growth in the manufacturing industry segment that develops and produces renewable energy technologies, with spillover for demand from traditional industry segments. Here we can note that the total number of direct and indirect jobs connected to the renewable energy sector in Germany was up from 160,000 in 2004 to around 381,600 by 2011 – the growth rate for green jobs was much higher than the rate of total employment growth (O'Sullivan et al. 2012). Extending renewable energy targets to the rest of the EU stimulated higher demand from German manufacturers.

The German government position also reflects the position of important national constituencies. The majority of the German population still demands ambitious climate policies and wants nuclear power to be replaced by renewable electricity. German industrial sectors have had mixed experiences from past policies, but major industrial groups have supported the EU 2030 package. The German Engineering Association (VDMA) and the German Renewable Energy Industry Association (BEE) have asked for climate *and* energy targets sufficiently ambitious to create stable opportunities for investment. The German electricity industry association has supported ambitious climate policies and a Market Stability Reserve despite higher costs for coal-based electricity. It has also supported renewable energy targets if combined with an EU-level market-based support system. This signals a compromise between the members that have faced big problems from the rapid expansion of renewable electricity and those that have made use of the new opportunities. The BDI, on the other hand, has called for a single GHG emissions reduction target, its level to be decided only if a satisfactory international agreement can be achieved at the 2015 Paris Summit– even though the most energy-intensive portion of the industry has been comprehensively shielded from the costs of the 2020 package. The German government has demanded the full continuation of carbon-leakage provisions under the 2030 package.

The European Council, with the German government, has also supported extension of the NER 300 mechanism to support innovative renewable energy and CCS projects (a new NER 400 mechanism) in the 2030 package. However, because of Germany's experiences of implementing CCS policies, the prospects for this technology option are poor there, even though coal-based power production still accounts for about half of total electricity production. The German government's push for speedy EU ETS reforms may be understood against this background.

With little hope of CCS becoming a short- or medium-term solution in Germany, the country will need stricter measures to replace coal with natural gas and renewables in its national electricity supply.

6.7 CONCLUSIONS

Germany experienced delays in implementation of the CCS Directive and the NIMs report under the EU ETS Directive. The German CCS transposition resulted in the cancellation of industrial development initiatives underway in this field of technology. Although the Renewable Energy Directive was transposed correctly and on time, there were challenges in implementing policies and measures in the areas of renewable heating and renewable transport fuels. All the same, Germany seemed well on the way to achieving the 2020 total renewable energy target, because of its massive investments in renewable electricity. Implementation here proved 'too successful', and reforms in 2014 aimed at slowing down the speed of investments and costs to consumers, which also made target achievement more challenging. As of 2014, Germany was not entirely on track as regards meeting its ESD target, due to problems with policies for the transport and building-stock sectors. Low ETS allowance prices have contributed to recent emissions increases also in the German ETS sectors. Thus, Germany now seems set to miss its self-imposed 40 per cent GHG emissions reduction target, unless planned additional policies can be put in place.

In essence, we can note a high degree of fit between the EU package and German policies. Poor fit and adaptation pressure from the EU thus explain little of the challenges and problems encountered as regards implementation. The EU package was not a game-changer – Germany succeeded in uploading much of what it preferred, and major parts of the package were readily absorbed with existing policies. There is no evidence that any major 'misfits' for Germany were created from negotiating a package of policies simultaneously.

Explanations for implementation problems can be found in domestic politics. Shifts in the German federal government help to explain implementation challenges regarding the EU ETS (EU-level implementation of backloading). The Federal Council (assembly of *Länder* governments with co-legislative powers) slowed down federal policy transposition for the CCS Directive and policymaking connected to target achievement under the ESD. Cost–benefit distributions among societal players affected implementation, spurring demands for reform of renewable energy policies in particular. Germany's consensual policy style enabled non-state actors to influence policies, variously facilitating and hampering the implementation of policy reforms. There were elements of horizontal and vertical fragmentation within the administrative apparatus that affected implementation of the package. Horizontal fragmentation caused delays in reforms of the ETS Directive and renewable energy policy; vertical fragmentation affected implementation of the CCS Directive.

Similar to the logic applied for creating agreement at the EU level, Germany applied various generous side-payments to redistribute costs and thereby foster acceptance for ambitious policies, among societal actors and the *Länder*. The comprehensive feed-in tariff system created costs but also benefits that attracted individuals, households, farmers and businesses alike to compensate for the costs caused by the system, resulting in the speedy on-the-ground implementation of renewable electricity. Implementation came with high cumulative costs to some actors, like the major utilities, which lobbied to have the feed-in tariff system scrapped in favour of a more market-adapted support system. This was supported by the Commission, which considered German policies incompatible with the internal energy market. Challenges in implementing the package served to spur new coordination and agreement between administrative units at the federal level that facilitated policy reforms aimed at slowing the renewable electricity boom and strengthening the EU ETS. Despite the re-packing of implementation policies, Germany continues to plan, at the national and EU levels, for a low-carbon energy system dominated by renewables by 2050. Germany has supported ambitious and binding goals as regards climate, renewable energy and energy efficiency for the EU 2030 package.

NOTES

1. One-third of total primary energy used in Germany in 2007 was supplied by oil, 25 per cent by coal and peat, 23 per cent by natural gas, and only around 10 per cent each for nuclear power and renewables. The electricity mix saw the share of coal at 49 per cent, natural gas at 12 per cent, nuclear power at 22 per cent, renewables at 14 per cent, and smaller shares of oil and waste.
2. Even by late 2013, the four major German utilities still owned only 7 per cent of national renewable electricity generation capacity (Energy Transition – the German Energiewende Blog, 29 October 2013).
3. The German Christian Democratic Party and its Bavarian sister-party, the Christian Social Union.
4. Specifically, the draft proposed that 'guarantees of origin' – the verification papers adopted in 2001 for control that electricity accounted as 'renewable' would actually be so – should be developed into tradable certificates (Toke 2008, p. 3).

REFERENCES

Agentur für Erneuerbaren Energien (2014), *92 Prozent der Deutschen wollen den Ausbau Erneuerbarer Energien*, accessed 15 May 2015 at http://www.unendlich-viel-energie.de/92-prozent-der-deutschen-wollen-den-ausbau-erneuerbarer-energien.

Boasson, E.L. and J. Wettestad (2013), *EU Climate Policy – Industry, Policy Interaction and External Environment*, Farnham: Ashgate.

Bulmer, S., D. Dolowitz, P. Humphreys and S. Padgett (2003), 'Electricity and telecommunications: fit for the European Union?', in K. Dyson and K.H. Goetz (eds), *Germany, Europe and the Politics of Constraint*, Oxford: Oxford University Press, pp. 251–69.

Deutsche Emissionshandelsstelle (2014), Press Background Paper for the 2013 VET Report, accessed 30 April 2015 at http://www.dehst.de/SharedDocs/Downloads/EN/Publications/2013_VET-Report_press_paper.pdf?__blob=publicationFile.

Duffield, J.S. and K. Westphal (2011), 'Germany and EU energy policy: conflicted champion of integration?', in V.L. Birchfield and J.S. Duffield (eds), *Toward a Common European Union Energy Policy*, Basingstoke: Palgrave Macmillan, pp. 169–86.

The Economist (2013), 'European utilities: how to lose half a trillion euros', 12 October accessed 5 April 2015 at http://www.economist.com/news/briefing/21587782-europes-electricity-providers-face-existential-threat-how-lose-half-trillion-euros.

Eichhammer, W., U. Boede, F. Gagelmann, E. Jochem, J. Schleich, B. Schlomann, J. Chesshire and H.-L. Ziesing (2001), *Greenhouse Gas Reductions in Germany and the UK – Coincidence or Policy Induced? An Analysis for International Climate Policy*, Karlsruhe: Fraunhofer Institute for Systems and Innovation Research; Brighton: SPRU; and Berlin: German Institute for Economic Research, accessed 30 April 2015 at http://www.isi.fraunhofer.de/isi-wAssets/docs/x/de/publikationen/wallfall/co2report-isi-e-final.pdf.

Eikeland, P.O. (2013), 'Electric power industry', in J.B. Skjærseth and P.O. Eikeland (eds), *Corporate Responses to EU Emissions Trading – Resistance, Innovation or Responsibility?*, Farnham: Ashgate, pp. 45–97.

Eikeland, P.O. (2014), *Implementing the EU 2020 Energy and Climate Package in Germany – Green Champion Struggling to Adapt*, FNI Report 9/2014, Lysaker: Fridtjof Nansen Institute.

ENDS (2008), 'Energy ministers welcome climate energy package', 28 February.

ENDS (2012), 'CO2 market intervention plan suffers new delay', 30 November.

Energy Transition – the German Energiewende Blog, 29 October 2013, 'Citizens own half of German renewable energy', accessed 13 February 2015 at http://energytransition.de/2013/10/citizens-own-half-of-german-renewables/.

EurActiv (2014), 'Berlin approves new incentives for electric car drivers', 26 September, accessed 30 September 2014 at http://www.euractiv.com/sections/transport/berlin-approves-new-incentives-electric-car-drivers-308700.

EurActiv (2015), 'Germany's CO2 emissions decrease for the first time in 3 years', 1 April, accessed 12 April 2015 at http://www.euractiv.com/sections/energy/germanys-co2-emissions-decrease-first-time-3-years-313430.

European Environment Agency (2011), *GHG Trends and Projections in Germany*, Germany_TP2011_country_profile.pdf, accessed 30 April 2015 at http://www.eea.europa.eu/themes/climate/ghg-country-profiles/greenhouse-gas-country-profiles.

European Environment Agency (2014), *Trends and Projections in Europe 2014 – Tracking Progress towards Europe's Climate and Energy Targets for 2020*, EEA Report No. 6/2014, Copenhagen: European Environment Agency.

European Environment Agency (2015), *EEA Greenhouse Gas – Data Viewer, Emissions Shares by Country EU28, 2012*, accessed 25 August 2015 at http://www.eea.europa.eu/data-and-maps/data/data-viewers/greenhouse-gases-viewer.

Federal Environment Ministry (2007), 'Costs and benefits of the German government's energy and climate package', October, accessed 9 October 2014 at http://www.bmu.de/fileadmin/bmu-import/files/english/pdf/application/pdf/bmu_hintergrund_iekp_en.pdf.

Federal Ministry for the Environment, Nature Conservation and Nuclear Safety (2011), 'The Federal Government's energy concept of 2010 and the transformation of the energy system of 2011', October, accessed 9 October 2014 at https://www.germany.info/contentblob/3043402/Daten/1097719/BMUBMWi_Energy_Concept_DD.pdf.

Federal Ministry for the Environment, Nature Conservation and Nuclear Safety (2013), *Renewable Energy Sources in Figures, National and International Development*, July, Berlin: BMU.

German Energy Blog (2015a), '18,948 electric passenger cars registered by end of 2014', 2 March, accessed 1 June 2015 at http://www.germanenergyblog.de/?p=18075#more-18075.

German Energy Blog (2015b), 'Climate levy for conventional power plants remains controversial despite softened proposal', 21 May, accessed 1 June 2015 at http://www.germanenergyblog.de/?p=18638.

Heine, D., J. Norregaard and I.W.H. Parry (2012), *Environmental Tax Reform: Principles from Theory and Practice to Date*, IMF Working Paper WP/12/180, Washington, DC: International Monetary Fund, accessed 23 October 2014 at https://www.imf.org/external/pubs/ft/wp/2012/wp12180.pdf.

IEA (undated), 'CO2 emissions from transport (million metric tons)', accessed 15 June 2015 at http://data.worldbank.org/indicator/EN.CO2.TRAN.MT.

IEA (1990), 'Germany balances for 1990', accessed 12 October 2014 at http://www.iea.org/statistics/statisticssearch/report/?&country=GERMANY&year=1990&product=Balances.

IEA (2007), 'German balances for 2007', accessed 12 October 2014 at http://www.iea.org/statistics/statisticssearch/report/?year=2007&country=GERMANY&product=Balances.

Jänicke, M. (2010), 'German climate change policy: political and economic leadership', in R. Wurzel and J. Connelly (eds), *The European Union as a Leader in International Climate Change Politics*, London: Routledge, pp. 129–46.

Karlseng, T.H. (2006), *Decreasing German Climate Ambitiousness Simply Due to Economic Problems, or Do Politics Matter?*, FNI Report 8/2006, Lysaker: Fridtjof Nansen Institute.

Katzenstein, P. (1987), *West Germany: Growth of a Semi-Sovereign State*, Philadelphia, PA: Temple University.

Lauber, V. and L. Mez (2004), 'Three decades of renewable electricity policies in Germany', *Energy & Environment*, 15 (4), 599–623.

O'Sullivan, M., D. Edler, T. Nieder, T. Rüther, U. Lehr and T. Peter (2012), *Gross Employment from Renewable Energy in Germany in 2011*, a first estimate at 14 March 2012, Research project commissioned by the Federal Ministry for the Environment, Nature Conservation and Nuclear Safety, accessed 13 April 2015 at http://www.germany.info/contentblob/3146650/Daten/3903428/BMU_GrosEmploymentRE2011_DD.pdf.

PV Magazine (2014), 'Oettinger hält EEG für nicht weiter reformierbar', 23 July, accessed 21 August 2014 at http://www.pv-magazine.de/nachrichten/details/beitrag/oettinger-hlt-eeg-fr-nicht-weiter-reformierbar_100016071/.

Schmidt, S. and R. Buehler (2007), 'The planning process in the U.S. and Germany: a comparative analysis', *International Planning Studies*, 12 (1), 55–75.

Spiegel Online (2009), 'Greenhouse gas sin tax: Germany joins EU in tying car fees to emissions', 27 January, accessed 8 August 2014 at http://www.spiegel.de/international/germany/greenhouse-gas-sin-tax-germany-joins-eu-in-tying-car-fees-to-emissions-a-603798.html.

Suck, A. (2005), 'The politics for a sustainable energy industry: renewable energy policy in the United Kingdom and in Germany', in D. Cohen and A. Héritier (eds), *Refining Regulatory Regimes: Utilities in Europe*, Cheltenham and Northampton, MA: Edward Elgar, pp. 147–82.

Toke, D. (2008), 'The EU Renewables Directive – what is the fuss about trading?', *Energy Policy*, 36 (8), 3001–8.

Verband der Automobilindustrie, *Annual Report 2014*, accessed 2 May 2015 at https://www.vda.de/en/services/Publications/Publication.~1225~.html.

Vitols, S. (1997), 'German industrial policy: an overview', *Industry and Innovation*, 4 (1), 15–36.

Die Welt (2011), 'Kohlendioxid soll unter die Erde kommen', 13 April, accessed 14 October 2014 at http://www.welt.de/wirtschaft/article13166077/Kohlendioxid-soll-unter-die-Erde-kommen.html.

Ziegler, N. (1989), *The State and Technological Advance: Political Efforts for Industrial Change in the Federal Republic of Germany, 1972–1986*, PhD thesis in Political Science, Harvard University.

7. Implementation in Poland

Jon Birger Skjærseth

7.1 INTRODUCTION

Poland initially accepted EU climate policy but has become considerably more reluctant over time. In 2007, Poland agreed to the EU 20–20–20 climate and energy targets in the European Council of heads of state and government. In December 2008, Poland also accepted the package of binding climate and energy policies for achieving these targets as a first step towards decarbonizing the EU by 2050. In October 2009, the European Council, including Poland, decided to support the EU goal of reducing GHG emissions by between 80 and 95 per cent by 2050 compared with 1990 levels (European Council 2009).

Since 2009, however, Poland has opposed new long-term EU climate policies. In 2011 and 2012, it vetoed the European Commission's (henceforth: Commission's) energy roadmap and the low-carbon roadmap that proposed a stepwise plan towards a decarbonized Europe by 2050. Polish opposition may weaken the content of new EU policies, because long-term climate and energy goals require unanimous agreement in the European Council. Poland is the EU's sixth largest member-state by population, and serves as an informal leader of the group of CEECs. The relationship between the EU and Poland is accordingly crucial to the development of long-term EU climate and energy policies.

Poland represents an interesting case as regards EU decarbonization efforts, as indigenous coal accounts for nearly 90 per cent of the country's electricity production and 50 per cent of its total CO_2 emissions. To what extent and how has Poland implemented the EU climate and energy package? How can implementation be explained, and what are the consequences for Poland's position on new long-term EU climate policies? Will new EU policies represent a 'Europeanization' of Polish climate and energy policy – or a 'Polonization' of EU climate and energy policy?[1]

7.2 POLISH CLIMATE AND ENERGY POLICIES UP TO 2007

Poland's domestic policies were dominated by the Soviet Union from 1945 to 1989. The key goal of state policy was industrialization, based on coal for power plants and heavy industry. Policymakers did not have to worry about the next election, and political decisions were taken independently of expert advice or societal concerns for the environment (Bokwa 2007). Critical reports on the state of the environment were not made public and were effectively excluded from public debate. Economic interests related to mining, energy, chemical and metallurgical industries served as the main 'link' between the state and society. Alliances representing these sectors could block any decisions that might threaten their interests (Bokwa 2007, p. 119).

From 1989, Poland embarked on the transformation from a centrally planned economy to a free-market system. This led to structural changes and modernization of the economy, followed by more effective energy use (OECD 2012). The largest energy-intensive industries, like steel and cement plants, were sold. Some unprofitable hard-coal mines were closed, and the entire energy industry was reformed and consolidated. After the turn of the millennium, the electric power industry was vertically consolidated into four groups consisting of many companies with a managing company for each group, partly owned by the state.[2]

In the 1990s, Poland started to prepare for EU membership by reforming its legal system. From then on, the EU has had significant influence on the country's energy and climate policies. The main energy objectives outlined were security of supply for Poland as an importer of gas from Russia, competitiveness and the environment. After lengthy deliberations within the government, the share of renewable energy sources was set to grow to 7.5 per cent in 2010 in the primary fuel mix, in line with the 2001 EU Directive on renewable electricity (Directive 2001/77/EC). In 2003, Poland's share of renewable energy was 5.1 per cent, based almost exclusively on combustible renewables: co-firing of coal and biomass and waste (IEA 2011); and 90 per cent of its electricity production was based on coal, representing nearly 60 per cent of its energy consumption (Eurostat 2007). Technology-neutral green certificates were introduced in 2005 to stimulate the cheapest renewable technologies. From 2005, plans for developing a Polish nuclear programme developed. Measures were also announced to promote energy-saving.

From late 2002, the CEECs were granted greater access to EU decision-making. Poland and the other CEECs were included as observers in the final stages of negotiations on the EU ETS Directive (Skjærseth and Wettestad 2008). In 2003, as part of its obligation under the UNFCCC, the Polish Ministry of the Environment presented the country's first climate action plan (Ministry of the Environment 2003). GHG emissions had dropped by 30 per cent since 1988 as a result of modernization, and a 40 per cent reduction target was adopted for 2020. The climate plan was rather vague as to practical implementation, however. The country had no real policy that would integrate climate policies into other national sector policies.

Poland joined the EU in 2004. It was sceptical of the already adopted EU ETS, and failed to get its NAP approved by the Commission before the ETS was launched in 2005. In 2005, the Commission provisionally accepted the Polish NAP, on the condition that allowances be slashed by 16.5 per cent. Polish industry (representing nearly 900 installations under the system) and the government reacted with strong indignation (ENDS 2005). In 2007, this pattern repeated itself with greater intensity when Poland prepared its NAP for the second trading period (2008–2012). Together with the Czech Republic, Estonia, Hungary and Slovakia, it took legal action against the Commission, but finally had to accept the Commission's initial downsizing proposal. Poland appeared unified in its opposition to lowering the number of allowances, arguing that the trading system could threaten its national economy (ENDS 2007). However, it was among the first to transpose the 2004 'Linking Directive' (Directive 2004/101/EC), which gives companies the right to purchase credits abroad through the Kyoto Protocol's flexible mechanisms, in order to comply with the EU ETS. Investments in cheaper external credits would ease the pressure on Polish industry.

7.3 NEGOTIATING THE PACKAGE: 2007–2008

The 20–20–20 targets were adopted by the European Council in March 2007. At that time, the Polish President was Lech Kaczynski from the Law and Justice Party (PiS). The EU targets were adopted by unanimity, but Poland, Hungary and the Czech Republic voiced concerns on how efforts to reach the climate target would be shared. These coal-dependent states also worried that a binding target on renewables could force them to invest in more costly energy sources (Skjærseth 2013; Ydersbond 2014).

In January 2008, the Commission formally proposed the climate and energy package. Poland was sceptical of the 20 per cent GHG reduction target adopted in 2007, to the EU ETS, and to the EU's ambition of leadership by unilateral action in the upcoming international climate negotiations. In Warsaw, the first government of Prime Minister Donald Tusk had taken office in October 2007; it sent somewhat mixed signals on renewables (Ancygier 2013). Poland was positive to co-firing biomass and coal, but was also concerned about increases in electricity prices. The stage was set for hard negotiations between the EU and Poland on the proposed package. Poland coordinated its position with the V-4, which from 1991 had come to consist of the Czech Republic, Hungary, Poland and Slovakia. When Poland took over the Presidency of the V-4 (July 2008–June 2009), one of the country's priorities was to 'promote exchanges of information on the preparation of positions concerning the energy and climate package' (Visegrad Group 2008/2009).

Negotiations on the package proposal intensified from July 2008, when France took over the EU Presidency from Slovenia (Skjærseth and Wettestad 2010). Poland fronted opposition to the 2005 baseline proposed by the Commission, and called for greater flexibility as regards achieving the 20 per cent GHG target, and for stronger safeguards against carbon leakage should the upcoming negotiations in Copenhagen fail. Poland and other CEECs preferred a 1990 package baseline that would benefit economies in transition, as their GHG emissions had dropped considerably after 1990. Poland also favoured three specific changes to the proposed revision of the EU ETS: price controls in the form of a carbon price ceiling; free allowances for electric power plants;[3] and greater financial assistance from auctioning revenues under the EU ETS.

Prior to the EU leaders' meeting in October 2008, concessions to Poland were prepared in the new context of the unfolding financial crisis. The Commission had proposed that 10 per cent of the revenues should go to the 'solidarity fund', whereas Poland was calling for 30 per cent. In November, the Baltic states joined with the V-4 alliance and teamed up in a coordination meeting in Warsaw (ENDS 2008b). The coalition of CEECs, headed by Poland, threatened to veto the whole package if the energy situation in these countries were not taken into account sufficiently (Ancygier 2013, p. 126). To prevent this, Commission representatives travelled to Warsaw to 'sell' the climate and energy package by emphasizing synergies and new low-carbon opportunities.

In December, Poland held a summit with French President Sarkozy. The meeting did not resolve the 'solidarity fund' issue, but a compromise was reached on granting power stations a longer derogation period from auctioning – from 2016 to 2020 (ENDS 2008c). The Polish position was that free allowances to protect energy-intensive industries exposed to international competition should apply to the electricity sector as well. This position was backed by a massive campaign at the national and EU levels, financed and fronted by Poland's conventional energy industry. That the energy industry was later praised by the government for its efforts underscores the unity between Polish business and the government in their ETS opposition (Ancygier 2013, p. 219).

The negotiations on the CCS proposal introduced significant changes in the measures for incentivizing CCS. The European Parliament also proposed CO_2 emissions limits on power stations, to force the use of CCS on future coal power – but this was opposed by Poland and other coal-dependent CEECs, foreseeing high CCS costs. After long and complex negotiations, it was agreed that 300 million allowances from the ETS NER 300 would be set aside to co-finance up to 12 commercial CCS demonstration projects and new renewable energy technologies.

With the EU RED, domestic actors were also engaged in influencing the Polish position. In August, the Polish electricity association PKEE responded to the Commission's proposals,[4] arguing that the package would damage the Polish economy by reducing GDP dramatically and raising electricity prices by 60 per cent (ENDS 2008a). Poland argued for a lower share than a 15 per cent increase in renewable energy sources (Ancygier 2013, p. 333), although this was put forward with significantly less intensity compared with its ETS derogations. For sectors outside the ETS included in the ESD, Poland accepted a generous national target of 14 per cent increase in GHG emissions by 2020 compared with 2005.

On 12 December 2008, a final compromise was reached on the EU ETS. First, Poland and other CEECs won concessions postponing the phase-in of auctioning for power plants. In 2013, certain stations would get up to 70 per cent of all allowances for free, decreasing towards zero by 2020. However, in exchange for free allocation, eligible installations would have to develop national plans to make their electricity sectors more efficient, and diversify their energy mix through investments worth at least the same as the value of the free allowances.

Second, as to solidarity, 88 per cent of the allowances to be auctioned each year would be distributed to the member-states. Ten per cent would be distributed to the poorer member-states, as proposed by the Commission. Two per cent would go as a bonus to member-states whose GHG emissions in 2005 were at least 20 per cent below their 1990 emissions: the CEECs. At least half of the auctioning revenues should be (that is to say, recommended – not a binding requirement) used to finance adaptation and mitigation. And finally, Poland did not get a carbon price ceiling.

The upshot was great variation in the distance between Polish policies and positions put forward with varying intensities, and the negotiated outcome. Poland strongly opposed a more ambitious ETS and managed to get several concessions. However, these concessions were far from meeting the Polish demands. On renewable energy sources, Poland was positive to co-firing biomass and coal, but had to accept a somewhat stricter target in the RED than preferred. Poland welcomed the CCS Directive and the NER 300, but successfully defeated an initiative to link CCS to emissions limits on power stations. The proposed ESD target for Poland – a 14 per cent increase in emissions – was seen as generous, and was accepted without resistance.

7.4 IMPLEMENTING THE PACKAGE

EU ETS

Poland failed to transpose the revised EU ETS (2009/29/EC) for the third phase (2013–2020) by 31 December 2012, as required. In September 2013, Poland (and Slovenia) received a Reasoned Opinion from the Commission (Commission MEMO 2013).

Article 10c of the Directive specifies the derogation from auctioning in the electric power sector. A draft list of power installations eligible for free allowances was issued by mid-2011 to meet two criteria: initiated before the end of 2008, and submission of an investment plan for clean technology equivalent to the market value of the free allowances. The list triggered fierce criticism from the environmental lobby, which claimed that it included fake 'phantom' power stations and that planned coal-power stations were listed as installations in the national investment plan for clean technology. In essence, Poland was blamed for using the derogations under the revised ETS to reinforce the dominance of its coal-fired power plants.

In 2012, a ruling by the Commission excluded 30 of the planned plants. According to the Commission: 'The Commission's decision also stipulates that certain investments proposed by Poland, primarily concerning new coal-based generation capacity, may not be used to justify the allocation of free allowances' (Commission MEMO 2012). This ruling, which Poland accepted, was linked to the fact that there would not be enough time to construct the plants within the third phase of the ETS, by 2020.

In spite of the derogations, there would still be revenues from auctioning, as only 70 per cent of the allowances could be distributed for free from 2013. Article 10 (3) of the revised ETS stipulates that member-states are to determine the use of the revenues generated – at least 50 per cent of these revenues should (again: recommended) be used for low-carbon activities. Poland decided that revenues should go to the general budget (ENDS 2013c). Transition to auctioning of allowances required a system to administer the auctioning process. Germany and the UK appointed their 'own' auctioning platforms, but Poland failed to specify an auctioning platform.[5]

Energy-intensive industries exposed to international competition and at risk of 'carbon leakage' will continue to receive free allowances until 2020 and beyond. Free allocation is to be carried out on the basis of product benchmarks that reward best practice in low-emissions production. Member-states must calculate the number of free allowances to be allocated to each installation in NIMs. Poland was one of the first member-states to submit its NIM in 2011. However, it opposed the benchmark rules for energy-intensive industry, arguing that the fuel mix should be taken into account. Poland took the case to court, where it lost in March 2013 (ENDS 2013b).

The Commission also designed state-aid rules for energy-intensive industries facing rises in indirect electricity costs under the ETS. Poland (and Germany) wanted the proposed list to be expanded, with compensation made available for all indirect costs (ENDS 2012a). The Commission had proposed 85 per cent compensation in 2013, falling to 75 per cent by 2020. In 2012, the state-aid rules were adopted, allowing 85 per cent compensation – in line with the original proposal (Commission Press Release 2012).

Thus, we see that Poland continued its systematic efforts to make the ETS fit with its coal-based electricity production after the adoption of the revised ETS Directive. Polish resistance to implementing the ETS proved generally unsuccessful: Poland lost in most cases, and had to accept the Commission's interpretation or the will of the qualified majority.

Effort-Sharing Decision

The ESD establishes differentiated annual national GHG emissions targets for the non-ETS sectors, 2005–2020. As a relatively poor EU country, Poland is allowed to increase its emissions by 14 per cent in sectors not covered by the ETS.

By 2013 Poland was clearly above the trajectory needed to meet the 2020 target (European Environment Agency 2014). The transport sector has been a major challenge. Between 1990 and 2009, the number of passenger cars per capita rose threefold and GHG emissions from the transport sector almost doubled in Poland (OECD 2012, p. 112). In 2008, the transport sector (mainly road transport) was the second-largest cause of Polish CO_2 emissions. Energy demand from the transport sector is expected to increase by more than 60 per cent by 2030 compared with the 2008 level (IEA 2011, p. 42). The Polish transport sector is affected by the EU Regulation (443/2009) on CO_2 emissions from new cars, but the country's climate policies for meeting the transport challenge are considered limited, too general or too weak (IEA 2011; OECD 2012). Poland's Energy Policy Plan 2030 does not include measures for the transport sector beyond biofuels as part of the RED (Ministry of Economy 2009).

Energy-efficiency measures in the residential and service sectors will also deliver contributions towards the EU target of 10 per cent reduction in the non-ETS sectors by 2020 (European Environment Agency 2013, p. 93). Nevertheless, the OECD recommends that the 'white certificate' scheme should be directly aimed at CO_2 reduction and not at energy savings (OECD 2012, p. 117). Because of the high use of coal for heating and hot water, and poor thermal insulation, residential and commercial buildings are responsible for 11 per cent of Poland's GHG emissions (OECD 2012). In 2008, Poland transposed the EU Directive on the Energy Performance of Buildings.[6] The Directive includes minimum energy performance standards for new and renovated buildings, and a system of energy performance certificates. According to the OECD (2012, p. 117), the energy performance certificate system should be extended to cover CO_2 emissions as well.

In summary, despite slow progress, Poland has projected 2020 ESD emissions below the 2020 annual targets under current national policies and measures. This optimism is not shared by all, however: concerns have been raised, particularly as regards the rising emissions in the transport sector. The most recent European Environment Agency evaluation shares this concern (European Environment Agency 2014). The ESD includes several flexibility mechanisms that will ensure goal attainment, including banking and borrowing emission budgets between years and buying AEAs from member-states that have over-achieved their targets. Member-states may also use external credits, as in the ETS.

Renewable Energy Directive

The RED includes a binding target for Poland to increase the share of renewable energy sources from 7.2 per cent in 2005 to 15 per cent in 2020 (of gross final energy consumption) along a trajectory of interim targets. For transport, the Directive requires that 10 per cent of energy use must come from biofuels or other renewables by 2020.[7] Legal transposition of the Renewable Energy Directive has had a bumpy ride in Poland. All EU member-states were to have transposed the Directive by 5 December 2010. In January 2011, Poland received a letter of formal notification, and in March 2012 came a Reasoned Opinion. In March 2013, the Commission referred Poland (and Cyprus) to the Court of Justice for failure to transpose the Renewable Energy Directive. The Commission proposed daily penalties based on the duration and gravity of the infringement (Commission Press Release 2013). Poland responded by adopting the 'Small Tri-Pack' Act of 26 July 2013, amending the Energy Act which entered into force in September that year (Miller Canfield 2013). The Act does not concern renewables specifically, but introduces amendments to the green certificate system. It is seen as a 'minimum' solution, and legal experts disagree as to whether it will prove sufficient for the Commission.[8] In June 2015, Poland received a Reasoned Opinion from the Commission about ensuring correct implementation of the Renewable Energy Directive, particularly concerning the target of 10 per cent share of renewables in transport by 2020 (Commission DG Energy 2015).

A recurrent issue has been whether new legislation should reduce support for co-firing biomass and coal and increase support for less mature technologies by means of feed-in tariffs. Nearly half of the financial support for renewable energy has been going to co-firing biomass with coal. The green certificate system is planned to continue until 2017, or 2020 at the latest. That means that a significant part of Poland's increase in renewable energy may prove short-lived if coal becomes cheaper than biomass.[9] Co-firing has contributed to oversupply of green certificates, low demand from energy retailers and a low value of the certificates (ENDS 2013a). In addition, co-firing does not contribute to development in renewable industries, and may possibly lead to deforestation. A second issue is whether a new system should stimulate large energy companies – or smaller local ones, as in Denmark and Germany.

As a result, the implementation process in Poland has been characterized by draft proposals, consultations, amendments and new draft proposals. The latest draft is aimed exclusively at the 2020 target; designed for large companies, it represents a continuation of 'dedicated' co-firing. As a result of the legal uncertainty, investments in windpower have slowed down (*Financial Times* 2013). Still, Poland saw its share of energy from renewable sources increase from 9.2 per cent in 2010 to 11.3 per cent in 2013. Poland may very well reach its lower 15 per cent target for 2020 due to significant co-firing biomass and coal. However, RED implementation is likely to stimulate mainly short-lived co-combustion of coal and biomass as well as some windpower by the big energy companies. That will provide limited opportunities for new business development and new low-carbon energy for the future.

CCS Directive

The CCS Directive (2009/31/EC) establishes a legal framework for safe geological storage of CO_2. Transposition of this Directive determines whether new modernized coal plants will be made CCS-ready. In 2011, the IEA advised Poland to transpose the CCS Directive swiftly, so as to be able to offer clarity to investors in the power sector (ENDS 2011).

Transposition of the CCS Directive in Poland has been a lengthy and somewhat confusing process. On the one hand, the government has taken CCS seriously and has had a genuine desire to make a workable legal framework (Jendroska 2014). Poland has favoured on-shore storage; and public opposition has been low compared with that in, for instance, Germany and the Netherlands. Poland has ample CO_2 storage capacity – estimated to 6–7 $GtCO_2$, with deep saline aquifers accounting for the largest share (IEA 2011). On the other hand, national legislation for transposing the CCS Directive for demonstration projects was not agreed until September 2013, more than two years after the deadline. Two regulations were still being drafted to enable full implementation of the CCS Directive.[10] The government has cancelled further work on a roadmap for CCS in Poland. Such a roadmap, identifying the country's storage potential, is needed to assist decisions on where new low-carbon power plants should be built in the future.

The Polish government (the Ministry of the Environment and the Ministry of Economy) aimed to build two large demonstration projects by 2015 as part of the wider EU CCS programme (IEA 2011). The first project was the PGE-owned Belchatow power plant, the largest coal plant in Europe, where post-combustion was initiated in 2009 on an 858 MW lignite ('brown coal')-fired unit under construction (PGE 2012). The second involved the hard-coal Kedzierzyn SA plant, where pre-combustion CCS was planned by ZAK and PKE, assuming 90 per cent capture. Both projects have now been cancelled, due mainly to lack of funding, but also because of legal barriers following the late and deficient transposition of the CCS Directive (ClientEarth 2013, p. 42). That has also contributed to problems with assessing the CCS-readiness of new coal-powered plants (ClientEarth 2013, p. 43).

Thus, Poland's late transposition of the CCS Directive has been a factor contributing to the failure of CCS-readiness assessment for new plants and for cancellation of CCS demonstration projects, although lack of funding is the main reason. The lack of funding is due in part to the low carbon price, which has weakened NER 300 and generally provided scant incentives for CCS investments. And so this low-carbon opportunity particularly tailored to coal plants and to countries like Poland is defunct, at least for the time being.

Conclusions

Assessed against the EU 2020 targets, Poland's implementation of the EU's climate and energy package may very well lead to goal attainment. However, assessed in terms of 2050 decarbonization, the picture looks quite different. Poland intends to continue with coal as its primary energy source in the foreseeable future. It also foresees a significant increase, especially in electricity and gas demand, by 2030. Such increases in demand are to be met by energy-efficiency improvements and by diversifying the energy mix. Poland plans to invest €25 billion in eight new coal plants, nuclear and liquefied natural gas (LNG) (ENDS, 22 November 2012). It has also been active in exploring shale gas, but only about 80 of 300 drillings needed to estimate the potential had been conducted as of 2014.

Diversification by renewables may prove short-lived if coal becomes cheaper than biomass. Late transposition of the CCS Directive and problems with financing have contributed to cancellation of all CCS pilots and the failure to make new plants CCS-ready. Finally, Poland lacks a specific climate policy, and that makes it difficult to curb rising emissions in the non-ETS sectors, transport in particular.

In essence, Poland has mainly opposed and reluctantly accepted the climate and energy package and then absorbed it, seeking to make it fit with existing policies and the country's energy mix. However, transposition and application of the various policies have not yet led to any significant changes in policies or behaviour.

7.5 EXPLAINING IMPLEMENTATION

The first explanation of implementation looks at variation in 'fit' and adaptation pressure between EU requirements and the Polish situation (see Chapter 2). The second explanation explores the significance of domestic politics for explaining implementation. Has Poland been able to reproduce EU-level synergies from the package approach domestically?

Fit and Adaptation Pressure

Good fit between EU requirements and the national situation – assessed in terms of the 'distance' between pre-existing policies, energy-economic situation, negotiating positions and the final negotiated EU outcome – is likely to promote implementation. For Poland, the 'fit' has varied significantly with the specific EU climate and energy policies. A nearly unified Poland has opposed the EU ETS ever since joining the EU, unsuccessfully challenging the Commission over several issues. It seems reasonable to conclude that the fit concerning the ETS was low, as the revised ETS deviated significantly from Polish policies and positions. The fit with the energy mix was also low. The EU average share of coal in gross inland energy consumption is 17 per cent, as against over 50 per cent in Poland in 2012 (Eurostat 2014). In the EU, about 30 per cent of electricity is produced from coal, compared with nearly 90 per cent in Poland (Commission DG Energy 2009). Polish electricity production is based mainly on indigenous coal, responsible for about 50 per cent of total CO_2 emissions.

As expected, low fit and high adaptation pressure have corresponded with strong opposition and efforts to absorb changes by implementing the ETS to make it suit coal-based electricity production. Indeed, the combination of low fit and high adaptation pressure resulting from the negotiations is a reasonable explanation of Poland's opposition and implementation challenges. Moreover, the failed EU ambition of an OECD-wide carbon market by 2015 could have eased the adaptation pressure for energy-intensive industries exposed to international competition.[11] However, relatively few industrial sectors are electro-intensive in Poland. The most exposed sectors are non-ferrous metals, iron and steel, and chemicals (Sartor and Spencer 2013).

Poland argued for a somewhat lower renewable energy share than that proposed by the Commission. However, it did not *insist,* and accepted the 15 per cent target by 2020 compared with a 7.9 per cent renewable energy level of gross final energy consumption in 2008 (IEA 2011). As Poland strongly opposed revision of the EU ETS, in the 2008 negotiations it had to prioritize its resources in terms of time, energy and political 'capital'.[12] Concessions to Poland on the ETS would make it more difficult to get concession on the Renewable Energy Directive, as the 27 member-states had to agree on the package in one round. Combined with an ongoing renewable energy policy based on technology-neutral certificates stimulating co-combustion of coal and biomass, existing Polish renewable energy policy was tailored to fit the interests of the large coal plants. Thus, we may say that the relationship between EU requirements and Poland's implementation of the RED is only partly in line with expectations. Despite only a moderate degree of misfit, Poland has experienced significant transposition and application problems.

Regarding CCS, the fit was high and the adaptation pressure low. There was a genuine willingness to develop CCS pilot projects based on a workable legislation, and public opposition was low. Here we see that implementation of the CCS Directive is not in line with expectations. Despite the high fit, Poland has experienced significant problems in transposing and applying the CCS Directive.

The ESD target fit well with Poland's lack of an ambitious climate policy. The country has not adopted a new climate policy since 2003. The ESD did not place much pressure on Poland, as new policies and measures were apparently not needed to achieve the target of 14 per cent increase in emissions. This shows how low pressure, no implementation and goal attainment may go hand in hand.

Taken together, the 'misfit' proposition outlined in Chapter 2 has proven difficult to assess due to the differing nature of the EU instruments and the time still remaining until 2020. Observations indicate that there is a systematic relationship between fit and adaptation pressure and implementation challenges – in line with expectations for the ETS and the ESD, partly for the RED and not for CCS. However, it is evident that negotiating a package of policies at the same time has led to varying fit and adaptation pressure. With the RED and ETS, there has been a trade-off in concessions, whereby a more ambitious RED target was accepted in order to boost the chances of gaining concessions on the ETS. Thus, we see that package policies can affect adaptation pressure for individual pieces of legislation.

Concerning the climate and energy package as a whole, one idea was for the EU to demonstrate its 'leadership by example' prior to the December 2009 climate Summit in Copenhagen. Poland has never shared this idea. Quite the contrary: it has consistently insisted that the EU should take into account the contributions from other major emitters before developing its own policy. Another idea was to strengthen EU energy security. Energy security is also a key concern for Poland, since the country imports 95 per cent of its crude-oil demand and two-thirds of its gas demand, mainly from Russia (IEA 2011). Even though Polish energy dependency is well below the EU average (Eurostat 2014), the Polish government has been using the energy-security argument to stick to indigenous coal, develop shale gas and build a nuclear industry. From the Polish perspective, an ambitious climate policy based on carbon-pricing and renewable energy subsidies could undermine coal and physical energy security, as well as leading to less affordable energy. Poland has never shared the EU's ambitions in climate policy, and this is reflected in the absence of a genuine climate policy.

The low fit between the EU and Poland concerning the package as a whole is roughly in line with observed opposition and implementation challenges. Still, additional explanations are needed, particularly for the RED and CCS, which have encountered implementation problems that are difficult to explain from the perspective of fit and adaptation pressure.

Domestic Politics

According to the above-mentioned perspective, and as explained in Chapter 2, variation in state responses to common EU legislation, or state responses to different EU legislation, can be traced back to the government itself, the society, the relationship between state and society, and administrative organization.

Governmental actors

The Polish government has changed since March 2007, when the European Council adopted the 20–20–20 targets. At that time, the government was headed by Lech Kaczynski, Chairman of the PiS. Kaczynski was a conservative climate sceptic, and his party has been characterized as 'climate deniers' (*Guardian* 2009). In November 2007 came the first government of the centre-right politician Donald Tusk, Chairman of the Civic Platform, which was somewhat less critical of climate policies. Tusk also secured a second government from 2011, and was in charge of the negotiation and implementation of the climate and energy package until autumn 2014. Tusk has been characterized as a strong leader who grants little autonomy to his ministers. These include the Minister of Economy, with responsibility for energy policy and the RED; and the Minister of Environment, with responsibility for climate policy and the ETS. With some exceptions, the interests and positions expressed by these ministers have shown a high level of continuity in supporting conventional coal and later nuclear energy (Ancygier 2013).

The majority government is based on the support of the agrarian Polish People's Party (PSL). In times of majority rule, only legislative proposals from the government get through the Parliament. This means that Tusk's government has had full authority over the transposition process since the package negotiations, within the limits set by EU law. Tusk has repeatedly engaged in Polish energy policy, viewing nuclear, coal and shale gas as the answer to the problems of energy dependency and the expected increase in demand for electricity, gas and oil. He has apparently told Polish media: 'The future of Polish energy is in brown and black coal, as well as shale gas' (Greenpeace 2013). As a response to the Russia–Ukrainian crisis in 2014, Tusk proposed a new EU energy union – with full exploitation of coal and shale gas resources in Europe as one of its pillars (EurActiv 2014). The crisis has reinforced Poland's belief in indigenous coal as the key to long-term energy security. Tusk accepts Poland's commitments from the climate and energy package to the extent that coal is taken into consideration.

In essence, then, we see that continuity in majority government, which holds veto power over transposition of laws, since November 2007 (before the EU package negotiations) has *not* served to promote implementation. Poland's change to a less climate-sceptical government after the adoption of the 20–20–20 targets can hardly explain the opposition and implementation challenges observed. There are virtually no political parties represented in the current Polish parliament that support the EU's long-term decarbonization vision (Bukowski 2013).

Affected societal actors

In Poland and in contrast to the EU level, electric-power producers and energy-intensive industry have apparently remained unified in their opposition to the climate and energy package since the 2008 negotiations. The Polish Chamber of Commerce – representing over 150 business organizations – has prepared an assessment of the EU's Energy 2050 roadmap, together with the largest power producers (EnergSys 2012). For the energy sector, electricity prices are expected to increase with implementation of the current 2020 package, and decarbonization policy will lead to significantly higher wholesale electricity prices beyond 2020. The annual costs for Poland will rise sharply, weakening the competitiveness of its industry, in turn leading to lower economic activity and higher unemployment. The impact on energy-intensive industry will also be negative as a result of the increase in electricity prices and carbon costs. Furthermore, households will also be negatively affected as a result of the increase in electricity and heat prices combined with decrease in income due to lower GDP growth.

As EU climate and energy policies are blamed for increase in electricity prices, these policies are unpopular among the Polish voting public. This must be understood in a context where the people have high expectations of improving their standards of living after the socialist years – and, to be re-elected, lawmakers must respond to and promote constituency interests. Studies prepared by the government have also tended to focus on the negative effects for high-carbon economies (see Boratyński et al. 2014).

The benefits identified are related to the development of low-emission technologies, but most of these (windpower stations, nuclear power stations, biogas-fired CHP plants) are and will remain based on imports. The renewables industry (other than co-firing) is dominated by foreign companies and depends on technology import (Polish Investment Agency 2013). Innovation and first-mover advantages in CCS, wind or solar are not considered a viable option. The only case where Polish companies can create added value of any scale is in biomass technologies. In essence, costs will be concentrated to the major economic sectors, whereas the benefits to those sectors and to society are judged negligible.

The four state-owned electric-power groups form a 'social bloc', particularly in their opposition to a new renewable energy law aimed at stimulating decentralized renewables. These groups guard their position, fearing what they see in Germany – where large German utilities have been outperformed by a renewable energy policy based on feed-in tariffs, and there has been a surge of new decentralized renewable power production followed by the shut-down of conventional power plants. No Polish political party has taken a clear position on limiting the role of coal in the economy (Bukowski 2013, p. 197).

Comprehensive and independent studies have argued that transition to a low-carbon economy in Poland will benefit investors and economic growth, reduce energy consumption, develop technology, create jobs, raise the level of energy security and improve health (see Bukowski 2013). However, Polish societal actors view 2020 policies as part of the EU's long-term low-carbon effort. This makes their *perceptions* of how policies will affect them in the long term the most important concern.

The EU package approach has not contributed to levelling out (perceived) costs between affected sectors in Poland. Carbon pricing and renewable energy subsidies have negative effects, especially on the coal-based electric-power sector. An attempt to transfer allowances from electric-power companies to energy-intensive industry failed (Ancygier 2013, p. 199). From the Polish perspective, its coal-based electric power industries face carbon pricing, renewable energy sources in which they have limited interest beyond co-firing, and an EU vision in which they will have no future in the absence of CCS. Energy-intensive industries fear that carbon pricing will raise their production costs and reduce their competitiveness. Finally, the general public is concerned about higher electricity prices and higher fuel expenses. This broad-based societal resistance clearly contributes to explaining Polish opposition to the EU ETS, and RED implementation challenges.

Policy style and access

Opposition from negatively affected societal actors does not necessarily mean *influence*. Most legislative proposals in Poland involve consultations with interest organizations. For instance, the draft Polish Energy Policy until 2030 underwent extensive public consultation, and the results from the consultation were published in a separate report (Ministry of Economy 2009). Draft EU laws are also made subject to routine consultation before transposition.

While the formal consultation system grants broad access for all affected actors, some actors have privileged informal access to decision-making. As regards the Renewable Energy Directive, the dominant conventional energy groups are 'core insiders' and have a decisive role in shaping energy policy. They were active at the EU and national levels during the negotiations on the package in 2008 and were later praised by the government (see section 7.3). These groups also represent 'segments' of society. In addition, they are protected by keeping other competing interests at arm's length. The first draft law on implementing RED was prepared in secret, to avoid lobbying from groups that wanted subsidies to decentralized renewable energy technologies other than co-firing coal and biomass (Ancygier 2013). Organizations representing various types of renewables industry have had limited access to decision-making. Organizations representing interests in 'new' renewables[13] have been supported by the green groups – which have only limited political influence.

With the establishment of the Renewable Energy Department in Poland's Ministry of Economy in 2012, these 'new' interests gained a new channel. This department was established specifically to oversee Poland's implementation of the Renewable Energy Directive. Now renewable industries were systematically consulted before the second draft of the law was published. This draft also came to include feed-in tariffs for photovoltaic power plants and a significant reduction in support for co-firing biomass and coal. However, the draft law was, as noted, altered to better fit the interests of conventional energy groups – so we should not overestimate the relative clout of the new Renewable Energy Department. It employs around 15 officials, as against some 1000 in the Ministry of Economy who deal with other energy sources.

Differentiated access for different societal actors is also evident in the consultation processes on the CCS Directive. Observations from the transposition process show that selected stakeholders were prioritized, such as trade unions and business organizations (Jendroska 2014).

Poland's distinct approach to regulation in climate and energy policy can be placed somewhere between an imposition and consensual approach, with privileged access for the conventional energy industry in particular. Other interests, like the new renewable energy industry and green NGOs, have been partly excluded from decision-making, which has strengthened the influence of the conventional energy industry. Lack of participation in the making of EU legislation can hardly explain the opposition of the conventional energy industry to implementing RED and the ETS Directive. Our main observation is that the societal actors most negatively affected have also been granted 'privileged' access and have had most influence on decision-making. Thus far, then, the package has basically served to cement old alliances opposed to implementation of the EU climate and energy package.

Administrative organization

We have seen that the principal ministers involved in implementation have been aligned as to the main direction of Polish climate and energy policy. Politicians in Poland tend to operate more independently from public officials than in most other EU countries. Accordingly, administrative organizations involved in executing policy may have wide room for manoeuvre when policies are to be implemented.

Poland's administrative energy system has been characterized as fragmented (IEA 2011). This organizational fragmentation is made worse by law fragmentation: the energy law has been amended some 50 times now (Ancygier 2013). Relevant environmental legislation is also complex – 20 legal Acts adopted by the parliament as well as about 100 executive regulations (Jendroska 2014). That being said, however, Polish decision-makers have acknowledged administrative fragmentation in climate and energy policy as a problem, and there have been many plans for improving coordination (IEA 2011, p. 23).

In 2009, the Ministry of Environment and the Ministry of Economy signed a new integrated strategy – the first time that representatives from these ministries had sat around the same table to develop a joint strategy. The energy/environment strategy was partly related to the EU climate and energy package. However, the low-emission economy plan under this strategy is extremely vague: 'It is assumed that the final effect of the National Programme will be a set of activities designed to directly or indirectly reduce greenhouse gas emissions' (Ministry of Environment 2013, p. 59). Efforts to transpose the RED, ETS and CCS Directives were undertaken by different administrative organizations at different points in time. Legislative and administrative fragmentation and weak coordination protract the transposition of EU law in Poland. In 2012 and 2013, Poland was cited for the most EU environmental infringements, after Spain, Italy and Greece (Commission DG Environment 2014).

The delayed transposition of the CCS Directive must be seen as mainly the result of administrative and legal fragmentation (Jendroska 2014). There was a genuine willingness to develop workable legislation, the 'fit' with coal was high and public opposition was low. The difficulties were multiplied by the decision to reform existing and fragmented legislation (rather than by adopting a separate legal Act) and by differing opinions as to whether transposition should be limited to demonstration plants only. The Ministry of the Environment held that this would not be possible under the CCS Directive – whereas the Ministry of Economy, supported by the Ministry of Foreign Affairs, claimed that it would be in compliance with the Directive, an interpretation later supported by the Commission.

EU-level policies are to be implemented partly at the local level. This creates a vertical link which can lead to disintegration of policies. The local level is particularly important concerning renewable energy, where administrative barriers in the form of fragmented laws are a key problem, especially those relating to land-development plans, environmental impact assessments and the licence requirement for businesses dealing with electricity generation (ClientEarth 2013). For Poland, the current and planned renewable energy law does not significantly stimulate local decentralized renewables compared with the case of, for instance, Denmark, although Poland has fewer wind resources.[14] However, a survey on renewable energy sources among municipal authorities indicated that municipalities are in direct opposition to the government's prioritizing of large-scale centralized power (Ancygier and Szulecki 2014).[15] A bill put forward by the PiS Party may further protract the development of local renewable energy by shifting the responsibility for building permits to the regional level. The Polish wind sector fears that this may bring wind-power into conflict with landscape protection banning large structures like turbines from being erected closer than 4 km from the nearest dwelling (ENDS 2014).

Although the situation has improved somewhat since 2009, horizontal administrative fragmentation remains a challenge for transposition. This helps to explain CCS implementation challenges in particular. As to the vertical dimension, local authorities are most relevant for implementation of the RED other than co-firing. Local administrative pressure in favour of more decentralized sources of renewable energy strengthens our observation that the large conventional energy groups are the main actors obstructing a more ambitious renewable energy policy. The significance of the package concerning horizontal fragmentation is that it has stimulated greater strategic coordination between the Ministry of Economy and the Ministry of the Environment. Attempts to transpose the RED, ETS and CCS Directives have been made largely in isolation.

7.6 CONSEQUENCES OF IMPLEMENTATION FOR NEW EU 2030 POLICIES

The combination of climate and energy as differently valued issues has apparently not promoted agreement on implementation or long-term decarbonization policies in Poland. The reason lies in the asymmetrical interests and values attached to energy security and climate change. Climate-change policies have few supporters in Poland – whereas energy security is seen as important, by society and policymakers alike. Policymakers view an ambitious climate policy as a potential threat to energy security to the extent it may threaten coal, shale-gas exploration or affordable electricity prices.

Concerning synergies, subsidies to promote renewable energy have only to a limited extent spurred 'green' growth in terms of new jobs as a result of co-firing biomass and coal as the preferred option. CCS might have become a new area for technological innovation and first-mover advantages, but all pilot projects have been cancelled. Concerns about air pollution may raise willingness to develop climate policy, particularly in coal-dependent countries. In Poland, however, air pollution is dealt with by end-of-pipe technologies, like the application of flue-gas desulphurization facilities (Council of Ministers, Poland 2008, p. 35). Moreover, the low carbon price, exemption from auctioning in the electric-power sector (70 per cent from 2013) and the decision not to earmark revenues from auctioning have prevented any effective impact from side-payments by linking revenues from auctioning of emission allowances to 'green' subsidies.[16] Opposition from energy-intensive industry has, however, been modified by the opportunity to compensate for indirect costs from the EU ETS. And so, issue-linkage mechanisms that promote EU-level agreement have only to a very limited extent been transformed into new low-carbon opportunities in Poland.

Poland was pressured to accept the EU climate and energy package. This is evident from the relatively low 'fit' between the EU requirements and Poland's pre-existing policies, its energy-economic situation and negotiating positions. Later resistance to the package and implementation of the EU ETS show that Poland was not happy with the outcome. Ironically, improved administrative coordination as a result of the package also made Poland more unified in its opposition. After the adoption of the EU climate and energy package, the Ministry of the Environment, the Ministry of Economy and the Ministry of Finance joined forces in a Centre for Climate Policy Analysis to assess the (negative) implications of EU climate and energy policy for Poland (World Bank 2011; Boratyński et al. 2014). Moreover, the fact that Poland lost most of its battles with the Commission created a bad atmosphere between Brussels and Warsaw as regards climate policy. In addition, the slow progress on a new international climate treaty has contributed to Polish opposition to new long-term EU climate policies.

The combination of poor experience from implementation and pressure to accept unwanted policies has led Poland to resist new EU climate and energy policies. Poland vetoed both the 2050 roadmap for moving towards a competitive low-carbon economy and the Energy Roadmap (see Chapter 5). In an online poll conducted just after the low-carbon roadmap veto, 91 per cent of the members of the general public surveyed approved the veto decision (Ancygier 2013, pp. 127, 182). Poland also opposed any interference in the carbon market, including 'backloading' of 900 million allowances proposed by the Commission. The backloading proposal was adopted in the autumn of 2013.

In January 2014, the Commission proposed a framework on climate and energy policies for 2030. Poland, supported by other CEECs, coordinated the opposition during the initial negotiations in the Council of Ministers and the European Council in March 2014. In May that year, Poland teamed up with the V-4 + Romania and Bulgaria and agreed on a common list of demands.[17] The major points were full national sovereignty over the energy mix, more EU subsidies to modernize energy systems, and a heavier burden on the 'rich' EU countries that were arguing for a more ambitious climate policy. Basically: those who demand a more ambitious climate policy should deliver the reductions themselves. As shown in Chapter 10, Poland had to accept a new EU 40 per cent GHG emissions reduction target before the 2015 Paris climate talks. On the other hand, the framework includes a range of concessions to Poland and other low-income member-states, with full national sovereignty over the energy mix, more subsidies to modernize energy systems (including coal) and a heavier burden on the 'rich' EU countries.

7.7 CONCLUSIONS

To what extent, and how, has Poland implemented the EU climate and energy package? We have noted significant transposition problems concerning the ETS, RED and the CCS Directives, and Poland's application record is not encouraging. Poland has tried to absorb the ETS to make it fit the big coal companies, and has not applied new policies and measures under the ESD. Nevertheless, Poland's emissions dropped significantly from 1990 as a result of modernization of the economy.

Poland intends to continue with coal as its primary energy source in the foreseeable future. Diversification by renewables may prove short-lived, because of implementation of the Renewable Energy Directive by mainly technology-neutral green certificates, favouring co-combustion of coal and biomass. The country lacks a specific climate policy, making it difficult to curb rising emissions in non-ETS sectors, transport in particular. In essence, Poland has opposed and absorbed the EU climate and energy package to make it fit with existing policies and its own energy mix, where coal accounts for 90 per cent of electricity production. The EU climate and energy package cannot be said to have been a 'game-changer' in transforming the energy system.

'Misfit' and adaptation pressure from the EU can explain Polish resistance to the revised EU ETS. The ETS was adopted before Poland entered the EU; moreover, carbon pricing is most costly for coal-dependent states. Adaptation pressures can also partly explain RED implementation challenges, as renewable energy sources (other than co-combustion) have never been particularly favoured by Polish governments. However, domestic political variables are also needed to explain Poland's opposition to the ETS, and its RED implementation challenges. The consistency in governmental priorities of coal, negatively affected societal actors and privileged access to decision-making for these actors are important. The main pressure has come from Poland's four state-owned energy groups, reinforced by low public demand for climate policy. Horizontal administrative fragmentation and lack of coordination have been a general challenge, regarding CCS in particular. Vertical distribution of competence to local levels is relevant mainly for the Renewable Energy Directive, where fragmentation of legislation represents a challenge to implementation.

As to the significance of a package approach, our first observation is that negotiating a package of EU policies adopted by unanimity tends to lead to varying fit and adaptation pressure, partly as a result of the give-and-take entailed in the various components of the package. This has affected the relative fit and pressure on the ETS and RED in Poland. Poland gained acceptance for only some of its demands and was promised significant benefits. Second, we have seen how implementation of a package affecting a broad range of sectors may expand the opportunities for sharing costs, including new alliances in policymaking and improving coordination among administrative organizations. However, these opportunities have been exploited only to a very limited extent in the implementation of the EU climate and energy package in Poland.

What are the consequences of the Polish position on long-term EU policies? Poland has increasingly resisted long-term EU policies by vetoing roadmaps for 2050. The reasons involve a combination of poor experience with implementation to date and unwanted EU policies. Poor experience can be traced back to lack of willingness, ability or opportunities to transform the linking of various policies and issues that promoted EU-level agreement to the national level: the EU package has not proven instrumental in combining different values and interests on energy security and climate mitigation. Quite the opposite: an ambitious climate policy is seen as potentially threatening for energy security to the extent it threatens Polish coal. The EU package has not spurred new 'green' jobs or energy technological innovation on a larger scale, or led to any significant synergies between air pollution and climate change.

Poland serves as an informal leader for other CEECs, and the EU member-states have been deeply divided over new climate and energy policies. The 2030 climate- and energy-policy framework adopted by the European Council in October 2014 reflects this situation. Is this a matter of Europeanization, or Polonization? Somewhere in-between – but further details will have to be settled before we can say where the EU is heading in practice.

In the meantime, we can refer to some conditions from the analytical framework to speculate as to what might drive Poland towards a 'greener' pathway in the future. Adaptation pressure from the EU is likely to grow for the ETS sectors. Expectations of a higher carbon price towards 2030 will make new investments in domestic coal increasingly risky. The low international price for hard coal is making domestic mines increasingly unprofitable. The EU pressure in favour of renewable energy sources will probably recede, but renewable energy – particularly solar – is getting cheaper and may become competitive without state subsidies. And domestic politics may change. Poland's new Prime Minister, Ewa Kopacz, has accepted the new EU domestic 40 per cent GHG reduction target by 2030, probably signalling a somewhat more 'climate-friendly' position than Donald Tusk. Social demands and organizations pushing for climate policies may become stronger. Public support for hard coal may decrease; moreover, local opposition to new lignite ('brown coal') mines is on the rise. This opposition to lignite is linked to local support for more decentralized renewable energy other than co-firing coal and biomass. Such pressure from 'below' may affect the priorities of (some) political parties on future climate and energy policies. Finally, lower political support and a stronger decentralized renewable energy industry may gradually reduce the political influence of Poland's major coal-based electric power companies.

NOTES

1. The concept 'Polonization' is borrowed from Ancygier (2013).
2. PGE, Tauron, ENEA and ENERGA. As regards PGE: 96 per cent of electricity generated from fossil fuels, mainly hard coal; Tauron: 86 per cent of electricity from hard coal and lignite; ENERGA: smallest, but about 25 per cent from renewables, partly hydroelectric. See Ancygier (2013).
3. Poland particularly opposed full auctioning from 2013.
4. PKEE is a member of Eurelectric; it comprises Poland's largest power companies, including ENERGA, PGE, Tauron and ENEA.
5. Poland has decided to use EEX, which is the official platform, on its behalf.
6. The Directive was revised in 2010 (2010/31/EU); in 2013, Poland was formally requested by the Commission to ensure full compliance.
7. The RED target for transport will not be analysed here. The transport sector will be analysed as part of the ESD.
8. Poland notified its full transposition of the Renewable Energy Directive on 29 January 2015.
9. Co-combustion is conducted mainly in outdated old boilers. These installations are covered by the Transitional National Plan for upgrading large combustion plants, which will remain in force until 2020, when biomass co-combustion is to be disconnected from the national energy system.
10. Email from the Ministry of Environment, Geology Department. On file with author.
11. Due to limited transmission capacity, the electric-power industry is not exposed to international competition beyond Europe.
12. This interpretation is also supported by Ancygier (2013).
13. Such as photovoltaics, and off-shore and on-shore windpower.
14. Danish windpower has been heavily subsidized and developed by encouraging wind-turbine cooperatives. This has secured high public acceptance for wind turbines, based on tax exemption for own family electricity generation and for purchasing shares in wind-turbine cooperatives, which invest in community windpower.
15. The survey is based on 262 online survey answers out of 2126 municipalities contacted.
16. However, Poland did establish a Green Investment Scheme funded from the sales of surplus assigned amount units as part of the Kyoto Protocol.
17. The Visegrad Group Countries, Romania and Bulgaria Joint Paper on the EU Climate and Energy Framework 2020–2030. May 2014. On file with author.

REFERENCES

Ancygier, A. (2013), *Misfit of Interests Instead of the 'Goodness of Fit'? Implementation of European Directives 2001/77/EC and 2009/28/EC in Poland*, Hamburg: Verlag Dr. Kovac GmbH.

Ancygier, A. and K. Szulecki (2014), *Does Local Energy Mean Renewable? Report from a Survey on the Acceptance for the Development of Renewable Energy Sources Among Polish Local Authorities*, ESPRi Report No. 1, May, Wroclaw: Environmental Studies and Policy Research Institute.

Bokwa, A. (2007), 'Climatic issues in Polish foreign policy', in P.G. Harris (ed.), *Europe and Global Climate Change: Politics, Foreign Policy and Regional Cooperation*, Cheltenham and Northampton, MA: Edward Elgar, pp. 113–38.

Boratyński, J., M. Zachłod-Jelec, M. Antoszewski, K. Wójtowicz, P. Sikora, W. Rabiega, R. Malarski and L. Kąsek (2014), *Economic Effects of the Proposed 2030 Climate and Energy Policy Framework on Poland and other EU Regions*, Warsaw: Centre for Climate Policy Analysis, April.

Bukowski, M. (ed.) (2013), *Low Emission Poland 2050*, Warsaw: WISE.

Chiavari, J. (2010), 'The legal framework for Carbon Capture and Storage in the EU', in S. Oberthür and M. Pallermarts (eds), *The New Climate Policies of the European Union*, Brussels: VUB Press, pp. 151–77.

ClientEarth (2013), *Implementation of EU Climate and Energy Law in Poland*, Warsaw: ClientEarth.

Commission DG Energy (2009), 'Europe's energy position', DG Energy, accessed 14 November 2014 at https://www.energy.eu/publications/KOAE 09001_002.pdf.

Commission DG Energy (2015), Poland asked to correctly apply the Renewable Energy Directive', DG Energy, April, accessed 4 June 2015 at http://ec.europa.eu/energy/node/2509.

Commission DG Environment (2014), 'Legal enforcement', accessed 3 November 2014 at http://ec.europa.eu/environment/legal/law/statistics.htm.

Commission MEMO (2012), MEMO, Emissions Trading: Questions and Answers on temporary free allocation of emission allowances for power plants beyond 2012. Brussels, 13 July, accessed 5 March 2013 at http://europa.eu/rapid/press-release_MEMO-12-562_en.htm.

Commission MEMO (2013), MEMO, September infringements package: main decision. Brussels, 26 September, accessed 4 March 2013 at http://europa.eu/rapid/press-release_MEMO-13-820_en.htm.

Commission Press Release (2012), State aid: Commission adopts rules on national support for industry electricity costs in context of the EU Emissions Trading Scheme. Brussels, IP/12/498, accessed 4 March 2014 at http://europa.eu/rapid/press-release_IP-12-498_en.htm.

Commission Press Release (2013), Renewable Energy: Commission refers Poland and Cyprus to Court for failing to transpose EU rules. Brussels, IP/13/259, accessed 13 March 2014 at http://europa.eu/rapid/press-release_IP-13-259_en.htm.

Council of Ministers, Poland (2008), *The National Environmental Policy Plan for 2009–2012, with its 2016 Outlook*, Warsaw.

ENDS (2005), 'Polish anger over EU NAP Decision', 9 March.

ENDS (2007), 'EU's tough line on carbon sparks eastern backlash', 1 August.

ENDS (2008a), 'EU climate plans "will damage Polish economy"', 4 August.

ENDS (2008b), 'Alliance against parts of EU climate plans widens', 6 November.

ENDS (2008c), 'France tables new EU carbon trade compromise', 8 December.

ENDS (2011), 'EC sheds light on CCS law implementation', 1 April.

ENDS (2012a), 'ETS state aid rules due in next few months', 10 April.

ENDS (2012b), 'Poland to invest eur25bn in energy infrastructure', 22 November.

ENDS (2013a), 'Poland mulls measures to fix green certificate scheme', 4 March.
ENDS (2013b), 'Poland loses legal battle over free CO2 allowances', 8 March.
ENDS (2013c), 'Germany's energy and climate fund in jeopardy', 3 June.
ENDS (2014), 'Change to Polish law could hinder wind power', 4 July.
EnergSys (2012), Polish Chamber of Commerce: *Assessment of the Impact of the Emission Reduction Goals Set in the EC Document 'Roadmap 2050' on the Energy System, Economic Growth, Industry and Households in Poland*, 15 February, Warsaw: EnergSys.
EurActiv (2014), *Poland Calls for EU Energy Union*, 2 April 2014, accessed 19 April 2014 at http://www.euractiv.com/sections/energy/poland-calls-eu-energy-union-301303.
European Council (2009), Presidency Conclusions, 29/30 October, Brussels, 1 December.
European Environment Agency (2013), *Trends and Projections in Europe 2013 – Tracking Progress Towards Europe's Climate and Energy Targets until 2020*, EEA Report No. 10/2013, Copenhagen: European Environment Agency.
European Environment Agency (2014), *Trends and Projections in Europe 2014 – Tracking Progress Towards Europe's Climate and Energy Targets until 2020*, EEA Report No. 6/2014, Copenhagen: European Environment Agency.
Eurostat (2007), *Energy: Yearly Statistics 2005*, accessed 30 April 2015 at http://ec.europa.eu/eurostat/documents/3217494/5613164/KS-PC-07-001-EN.PDF/a9aa88dc-59c7-40fc-9081-db5f7fb44813?version=1.0.
Eurostat (2014), *Energy Balance Sheets, 2011–2012*, Eurostat Statistical Books, accessed 30 April 2015 at http://ec.europa.eu/eurostat/documents/3217494/5785109/KS-EN-14-001-EN.PDF/16c0ac97-7dd6-4694-b22d-e77a36cb4e86?version=1.0.
Financial Times (2013), 'Polish renewables: investors await legislation details to see which way the wind blows', 22 May, accessed 25 June 2014 at http://www.ft.com/intl/cms/s/0/1fc54398-b703-11e2-a249-00144feabdc0.html#axzz3od4CPugL.
Greenpeace (2013), 'Polish public's view on renewables poles apart from government policies', accessed 9 April 2014 at http://www.greenpeace.org.uk/newsdesk/energy/analysis/polish-publics-views-renewables-poles-apart-government-policies.
Guardian (2009), 'Anti-gay, climate change deniers: meet David Cameron's new friends', 3 June, accessed 18 September 2014 at http://www.theguardian.com/politics/2009/jun/02/david-cameron-alliance-polish-nationalists.
IEA (2011), *Energy Policies of IEA Countries: Poland*, Paris: IEA.
Jendroska, J. (2014), *Carbon Capture in Poland: The Transposition of Directive 2009/32/EC into Polish Law*, London: University College London.
Miller Canfield (2013), 'Small power Tripack becomes effective', Globe Business Publishing, accessed 13 March 2014 at http://www.lexology.com/library/detail.aspx?g=6c994a3d-1f4e-4419-b452-7b4ba832be01%20and%20http://www.clientearth.org/reports/130816-climate-and-energy-clientearth-small-tripack-legal-analysis-eng-final.pdf.
Ministry of Economy (2009), *Energy Policy of Poland until 2030*, 10 November, Warsaw: Ministry of Economy.

Ministry of the Environment (2003), *Poland's Climate Policy: The Strategies for Greenhouse Gas Emission Reductions in Poland until 2020*, October, Warsaw: Ministry of the Environment.

Ministry of the Environment (2013), *Sixth National Communication to UNFCCC*, Warsaw: Ministry of the Environment.

OECD (2012), *Economic Surveys: Poland*, Paris: OECD.

PGE (2012), 'CCS demonstration plant', accessed 12 March 2014 at https://pgegiek.pl/index.php/ccs/ccs-demonstration-plant/.

Polish Investment Agency (2013), 'Renewable energy', Polish Information and Foreign Investment Agency, accessed 9 April 2014 at http://www.paiz.gov.pl/sectors/renewable_energy.

Sartor, O. and T. Spencer (2013), *An Empirical Assessment of the Risk of Carbon Leakage in Poland*, Working Papers 08/2013, Paris: IDDRI.

Skjærseth, J.B. (2013), *Unpacking the EU Climate and Energy Package: Causes, Content, Consequences*, FNI Report 2/2013, Lysaker: Fridtjof Nansen Institute.

Skjærseth, J.B. and J. Wettestad (2008), *EU Emissions Trading: Initiation, Decision-Making and Implementation*, Aldershot: Ashgate.

Skjærseth, J.B. and J. Wettestad (2010), 'Fixing the EU Emissions Trading System? Understanding the post-2012 changes', *Global Environmental Politics*, 10 (4), 101–23.

Visegrad Group (2008/2009), 'Visegrad group programme of the polish presidency (July 2008–2009)', accessed 30 January 2014 at http://www.visegradgroup.eu/documents/presidency-programs/2008-2009-polish-110412.

World Bank (2011), *Transition to a Low-Emission Economy in Poland*, Washington, DC: World Bank.

Ydersbond, I.M. (2014), 'Multilevel "venue shopping": the case of the EU's Renewables Directive', *Interest Groups & Advocacy*, 3 (1), 30–58.

8. Implementation in the Netherlands

Lars H. Gulbrandsen and Jon Birger Skjærseth

8.1 INTRODUCTION

The Netherlands supported the EU climate and energy package and seemed well prepared for implementing it. In 2007, the government established objectives similar to and beyond the 20–20–20 targets established by the European Council – including to reduce GHG emissions by 30 per cent from the 1990 level by 2020 – and aimed at taking a 'big step' in the transition towards a more sustainable energy system by 2020.

The Netherlands has traditionally been seen as an environmental frontrunner in the EU (Andersen and Liefferink 1997; Liefferink 1997), although somewhat less so in the case of climate policy. Our point of departure for assessing Dutch implementation was the expectation that a good match between the EU package and national climate- and energy-policy priorities would facilitate implementation of the package as a first step towards a Dutch low-carbon economy. However, this does not seem to hold for the Netherlands, where the energy mix remains dominated by fossil fuels and the renewable electricity sector has been developing slowly.

In this chapter we ask: Why does the Netherlands have trouble demonstrating significant progress in implementing an EU package that it supported and that was apparently in line with national policies? This question calls for a detailed assessment of Dutch implementation of the EU package to date with a view to long-term decarbonization.

8.2 DUTCH CLIMATE AND ENERGY POLICIES UP TO 2007

The Dutch economy is an open, trade-intensive economy, which has more than doubled in terms of economic activity since 1975. However,

GHG emissions fell about 9 per cent between 1990 and 2012 (European Environment Agency 2014), thanks to improved energy efficiency and a high share of natural gas that made Dutch carbon intensity relatively low compared with other high-income OECD countries.

The Dutch energy supply mix is dominated by fossil fuels, which represent more than 90 per cent of total primary energy supply. In 2012, natural gas supplied 42 per cent of energy, followed by oil (39 per cent) and coal (10 per cent) (IEA 2014). Natural gas has been the backbone of Dutch energy provision since the 1960s and dominates the fuel mix in electricity production (50 to 60 per cent over the past 30 years). In the 1990s, renewables assumed a prominent place on the policy agenda due to growing concerns about climate change. Dutch governments have used a range of instruments to stimulate renewable energy use, targeted mainly towards wind and biomass use, but have struggled to increase the renewables share in the energy mix.

In 1989, the country's first National Environmental Policy Plan set climate change as one of the key environmental topics to be addressed. As an EU member-state, the Netherlands had to implement EU climate policies and comply with the 1997 Kyoto Protocol. Under the EU burden-sharing agreement for the Protocol, the Netherlands was required to reduce its GHG emissions by 6 per cent over the period 2008–2012 compared with 1990 levels.

The first National Environmental Policy Plan was followed up with various policy instruments and agreements aimed at reducing the dependency on fossil fuels and improving energy efficiency. The Dutch government's strategy for meeting its obligations under the Kyoto Protocol was established in the National Climate Policy Implementation Plan (NCPIP), issued in two parts in 1999 and 2000. According to the NCPIP, half of the Dutch commitment is to be met through domestic policies, and the other half through use of the flexibility mechanisms under the Kyoto Protocol.

In 1997, the government introduced an energy investment tax deduction for companies, to stimulate renewable energy and energy efficiency. In 1999, after the liberalization of the electricity market, the scheme was replaced with an energy tax exemption for energy distribution companies. A system for trade in green certificates was also introduced, to ensure the authenticity of renewable energy supply.

Annual government expenditure on Dutch climate and energy policy increased sharply in the late 1990s (Faber et al. 2012). The next decade saw many attempts to switch to a low-carbon energy system while also improving security of supply and maintaining affordability. In 2003, a feed-in premium scheme replaced the energy tax exemption. This scheme primarily stimulated co-generation or combined heat and power (CHP) and co-firing of biomass in fairly large coal-fired plants, making the Netherlands a European and global leader in use of this technology. However, it did little to stimulate technological innovation and the transition to a more climate-friendly energy system in small and medium-sized enterprises. The scheme was terminated in 2006 because of budget constraints.

The Netherlands was one of the few member-states that actively supported an emissions-trading system in the EU, with plans already made for a domestic system (Skjærseth and Wettestad 2008). The Netherlands was also a frontrunner in developing CCS technologies and pilot projects. It started CCS research and development activities in 1988 – among the first countries in the world to do so – but most important in giving direction to Dutch interest in a legal framework for CCS were the discussions of an EU CCS Directive (Ecofys 2015).

Climate policies were given higher priority in the Netherlands after the Kyoto Protocol entered into force in 2005 and in the run-up to the first Kyoto commitment period, 2008–2012 (Faber et al. 2012). However, Dutch dependency on fossil fuels did not decrease, energy consumption continued to rise (mostly supplied by natural gas and coal) and the share of renewables in the total energy mix remained low. One explanation for the slow development of renewable energy and new technologies might be the lack of stability and long-term vision in Dutch climate and energy policies (Rathmann et al. 2010, p. 61).

The centre-left government in office from 2007 re-shuffled ministerial responsibilities for climate and energy policies. The Ministry of Housing, Spatial Planning and the Environment (VROM) retained primary responsibility for the overall climate, energy efficiency and renewable energy goals. But every sector ministry was given responsibility for implementation of the climate policies in its own sector. In addition to this interministerial division of responsibilities for meeting domestic targets, the central government signed an agreement with the regional and local governments. While the regional and local governments retain authority to establish their own climate-policy ambitions, these ambitions are outlined in light of the new, national climate targets in this agreement. In line with Dutch tradition, a key component of the climate plan was also the signing of agreements ('sustainability accords') with all key sectors of the economy, in order to bring target groups on board.

8.3 NEGOTIATING THE PACKAGE: 2007–2008

In 2007, the new centre-left government (the fourth Balkenende cabinet) clearly had ambitions of being a European frontrunner in the transition towards a more climate-friendly and sustainable energy system by the year 2020.[1] Partly in response to the EU 20–20–20 targets adopted by the European Council in March 2007, the government in December 2007 proposed the programme *Clean and Efficient* (*Schoon en Zuinig*), with four primary targets:

- to reduce GHG emissions by 30 per cent from the 1990 level by 2020;
- to increase the share of renewables in the energy mix to 20 per cent by 2020;
- to achieve energy-efficiency improvements of 20 per cent by 2020;
- to make a 'big step' in the transition towards a more sustainable energy system by 2020.

The targets signalled new ambitions for climate policy. They were clearly influenced by the EU 2020 targets, but indicated that the Netherlands would move faster in reducing GHG emissions. Synergies between EU-level and Dutch climate and energy policy were debated before and just after the EU package was proposed. The Netherlands supported the main structure of this package, with reduction targets for ETS and non-ETS sectors, and differential national targets in the ESD and RED. From the Dutch perspective, an ambitious and centralized EU climate policy based mainly on the EU ETS would be advantageous.

Two Dutch priorities were indicated for negotiating the EU package: competitiveness, and stringent/robust sustainability criteria for biofuels. Competitiveness surfaced as a Dutch demand for more free ETS allowances from energy-intensive industries, to reduce the risk of their moving out of Europe. This demand became more important with the emerging global economic crisis and rising concerns about the costs of the package. The Netherlands, together with Germany and the UK, prepared an informal proposal in the summer of 2008, setting out detailed criteria for identifying industrial sectors at risk of carbon leakage. The proposal, based on a quantitative assessment of increases in product prices and exposure to international trade, was initially well received by member-state governments and MEPs, and the European Commission (hereafter: Commission) applied the criteria for the first analysis of exposed sectors in September (ENDS 2008b). The proposal was later modified by the inclusion of qualitative assessments; most energy-intensive industry sectors were granted free allowances based on benchmarks.

The negotiations on the CCS proposal introduced significant changes in how to incentivize CCS. For the Netherlands, CCS had become a key priority: it was the only country besides the UK that supported setting aside 500 million allowances from the ETS New Entrants Reserve to co-finance CCS demonstration plants (ENDS 2008d). After lengthy negotiations, 300 million allowances from the New Entrants Reserve were earmarked for co-financing up to 12 CCS demonstration projects and other renewable technologies. The Netherlands and the UK argued for less stringent geological storage provisions in the CCS Directive, but this was opposed by Germany and the Commission (European Council 2008).

Throughout the negotiations on the RED proposal, both the national targets and the 10 per cent minimum target for renewable energy in the transport sector were upheld by the EU. However, several details as to how these targets were to be calculated were changed (Howes 2010). For example, only biofuels that met sustainability criteria would count towards the target. This was in line with the Dutch position. Besides pushing for stringent biofuel criteria throughout the negotiations, the Netherlands was apparently sceptical of the Renewable Energy Directive – together with France, it was actually accused of undermining the Directive (ENDS 2008c). The Dutch environmental assessment agency PBL (Planbureau voor de Leefomgeving; formerly MNP) followed up by warning that the Netherlands would not be able to attain the renewable target due to shortcomings with the new feed-in premium system (ENDS 2008e).

The Netherlands also paid significant attention to the link between international and EU climate policies because of its concerns for the competitiveness of Dutch industries in the absence of ambitious international commitments. Early in the negotiations, the Netherlands argued – unsuccessfully – that effort-sharing should be specified in advance if the EU decided to raise the ambition level to 30 per cent (ENDS 2008a). The prospects of an ambitious international agreement were directly related to the EU's climate target (20 per cent cut in emissions, to be changed to 30 per cent if other developed countries followed suit) and revision of the EU ETS with regard to allocation methods.

To summarize, the Netherlands supported the EU package and, although the Renewable Energy Directive was not particularly favoured, Dutch positions during the negotiations matched most of the EU outcomes well. Moreover, Dutch climate and energy policies presented in 2007 appeared largely compatible with the EU package. Hence, we would expect any pressure on the Netherlands to change policies significantly to have been rather low, and that Dutch implementation of the EU package would be fairly straightforward.

8.4 IMPLEMENTING THE PACKAGE

EU ETS

The Netherlands has transposed the revised ETS Directive fully, but there was a delay in the timetable for implementation. By September 2012, all the necessary legal documents had been amended and had entered into force. The ministry in charge was the Ministry of Infrastructure and the Environment.

The Netherlands has 447 installations covered by the EU ETS, representing just over 40 per cent of total GHG emissions in the country (European Environment Agency 2013). All the major international oil companies, with Royal Dutch Shell in the lead, operate in the Netherlands and are among the biggest allowance-holders.

For the third trading period, which began in 2013, energy-intensive industry exposed to international competition and at risk of 'carbon leakage' will continue to receive free allowances until 2020 and beyond. Member-states must calculate the number of free allowances based on benchmarks allocated to each installation in NIMs, which was published by the Ministry in January 2012 – thus overstepping the EU deadline.

Article 10 (3) of the revised ETS stipulates that member-states are to determine the use of the revenues generated – at least 50 per cent of these revenues should be used in low-carbon activities. The Dutch legal framework did not propose any binding rule on the use of auction revenues, but the government has still earmarked 100 per cent of the ETS auctioning revenues for domestic climate- and energy-related purposes (Commission 2014).

In September 2013, a society-wide Energy Agreement for Sustainable Growth (the 'Energy Agreement') was adopted by 40 Dutch organizations and institutions, including central, regional and local government; employers' associations and unions; nature conservation and environmental organizations; and other civil society organizations and financial institutions. It was initiated by the Social and Economic Council (SER), which is an advisory and consultative body of employers' representatives, union representatives and independent experts with the explicit aim of helping to create societal consensus on national and international socioeconomic issues. The 2013 Energy Agreement is a comprehensive climate and energy deal focused on achieving the climate and energy objectives for 2020 and committing to joint lobbying for making a properly functioning EU ETS a crucial factor in the long-term transition to a sustainable energy system. The parties have agreed to lobby for implementation, by 2020, of the following changes to the EU ETS:

- tightening up of the reduction path for the ETS cap, aimed at achieving the long-term goal of an 80 to 95 per cent reduction in GHG emissions for the whole economy by 2050;
- securing the position of companies exposed to international competition and the challenge of 'carbon leakage' by means of a 100 per cent free allocation of permits based on realistic benchmarks and actual production, based on the best performance in the sector;
- compensation for indirect (electricity) costs, based on best performance in the sector (Energy Agreement 2013).

Hence, while the Netherlands wants to tighten the annual cap on emissions from ETS sectors post-2020 and expand the ETS system internationally to promote cost-efficient CO_2 reductions, it also aims to secure the position of its energy-intensive companies exposed to international competition.

Effort-Sharing Decision

For installations outside the ETS sectors, the national target for the Netherlands under the ESD is a 16 per cent reduction in GHG emissions from 1990 levels by 2020. The non-ETS sectors in the Netherlands represented 55 per cent of GHG emissions in 2011. This included transport, housing, waste disposal, agriculture and forestry, aquaculture and some areas of industry. In 2013, non-ETS emissions were well below the Dutch 2013 ESD target (European Environment Agency 2014, p. 51), and implementation of currently planned additional measures is projected to reduce 2020 emissions below the target level (European Environment Agency 2014, p. 52).

Regarding domestic mitigation measures, the Dutch government in 2011 established a *Strategy on Climate Policy En Route to 2020* (Ministry of Infrastructure and the Environment 2011), which included policy measures in all non-ETS sectors. Sectoral CO_2 emission ceilings for 2020 have been established in the following non-ETS sectors: industry, transport, buildings, agriculture and horticulture. Non-CO_2

emission ceilings have been established for GHG from agriculture and other non-ETS sectors.

In 2011 the Dutch government also prepared a Sustainability Agenda (*Agenda Duurzaamheid*) advocating 'green growth' through such measures as smart and efficient regulations and improved financial incentives. Two key policy instruments – the National Roadmap Climate 2050 and the Local Climate Agenda 2011–2014 – are intended to assist the achievement of the GHG reduction targets for 2020 and beyond. These policy instruments focus on sustainable buildings, mobility, companies, production and climate-neutral cities/regions. Other priorities include decentralized sustainable energy production, maintaining natural gas production, increased energy efficiency in the buildings sector and the promotion of electric vehicles (European Environment Agency 2013).

The use of covenants – negotiated agreements between the government and various stakeholders – is a typical and important feature of Dutch climate and energy policy, as well as in other policy areas. The government has entered into such negotiated agreements with various stakeholders on, for example, energy saving, local generation of renewable energy and sustainable transportation. In 2011, the government introduced voluntary 'Green Deals' to promote energy efficiency and energy saving. The aim is to remove some of the barriers – like confusion about licences, lack of collaborative partners and ambiguous regulations – that might make stakeholders reluctant to invest in energy-saving and sustainable energy use. The Environmental Management Act also contains several core instruments relevant for reducing GHG emissions (Peeters 2011, p. 128).

In addition, there are currently two types of long-term agreements in place, aimed at ETS and non-ETS enterprises. Agreements aimed at non-ETS companies are referred to as LTA3 (Long-Term Agreement on Energy Efficiency for non-ETS enterprises). These cover the period 2001–2020 for large and medium-sized companies and entities in industry, agriculture and the service sectors. They aim at achieving a 30 per cent improvement in energy efficiency between 2005 and 2020. In turn, the government has agreed not to impose additional specific national measures aimed at energy conservation or CO_2 emission reductions on companies participating in LTA3.

The Netherlands has adopted ambitious objectives for reducing CO_2 emissions in the transport sector. In 2012, transport accounted for almost 19 per cent of total energy-related CO_2 emissions (IEA 2014, p. 40). On the basis of the Energy Agreement (September 2013) and the Climate Agenda for 2030 (Ministry of Infrastructure and the Environment 2013), the Netherlands has reaffirmed its ambition to reduce CO_2 emissions in the transport and mobility sectors by 17 per cent by 2030 and by 60 per cent by 2050 compared with 1990 levels. The most important policy measures implemented so far are fiscal measures, electrification in the transport sector, and initiatives encouraging biofuel blending and sustainable mobility. Average CO_2 emissions from new cars sold in the Netherlands were around 120 g/km in 2012, placing the country among the top three EU member-states in terms of carbon-efficient cars (European Environment Agency 2013). Here the Netherlands benefits from Regulation (EU) No. 443/2009, which requires car manufacturers to reduce CO_2 emissions from new cars, as explained in Chapter 5.

To summarize, the Netherlands replaced the 30 per cent GHG reduction target from 2007 with a 16 per cent target for non-ETS installations only, because of the EU ETS and the ESD. In spite of the financial crisis (which made it much easier to meet the GHG target), it is expected that the Netherlands will be able to reach the new target only with the help of planned additional measures. At present, domestic measures appear overlapping and fragmented, with various programmes partly affecting the same sectors. The frequent use of voluntary agreements adds a layer of complexity and makes it more difficult to predict whether the Netherlands will meet its target.

Renewable Energy Directive

The Dutch centre-right coalition minority government that took office in 2010 (the first Rutte government) lowered the national target for the share of renewable energy from 20 per cent of *primary energy* (the total energy demand of a country) by 2020 to 14 per cent of *final energy consumption* (the total energy consumed by end users), the EU-set target for the Netherlands under the RED. The second Rutte coalition government, formed in 2012, raised the target to 16 per cent by 2020,[2] but later moved the target date to 2023, in line with the negotiated 2013 Energy Agreement.

The Netherlands has transposed the RED fully, with only a minor delay in the timetable for implementation. By March 2011, all the necessary legal documents had been amended and had entered into force. The ministry responsible was the Ministry of Economic Affairs.

With the adoption of the EU climate and energy package, the 2010 centre-right coalition government (the first Rutte cabinet) announced that it aimed to meet the Dutch renewable energy target primarily through a new feed-in premium scheme, called SDE-plus, to replace the SDE arrangement (*Stimulering Duurzame Energieproductie*) introduced in 2008. According to the government, the SDE-plus scheme focuses solely on achieving the binding renewable energy target of 14 per cent by 2020. This is a technology-neutral and market-based scheme that promotes competition among technologies and projects. In 2012, the following technologies were covered by the SDE-plus scheme: wind energy (on- and off-shore), solar PV (certain types of applications), waste incineration, biomass fermentation, incineration and thermal conversion, hydropower, geothermal energy and solar thermal energy. Unlike the previous SDE scheme, SDE-plus is financed through a surcharge on the energy bills of households and industry, making it less vulnerable to changes in government budgets and priorities.

Most projects funded by the SDE-plus scheme are either biomass projects (but not biomass co-firing projects) or on-shore wind projects, reflecting their cost-effectiveness. Renewable energy technologies such as off-shore wind, tidal and wave energy, and solar tend to be more expensive than the maximum cost price under SDE-plus. However, these technologies may compete for funding in the 'free category', which encourages innovative options, because individual projects may be less expensive than the average cost price for off-shore wind, tidal and wave energy or solar (if they have access to additional financing, for example).

The Dutch National Renewable Energy Action Plan describes how the Netherlands intends to achieve the 14 per cent target for 2020 (Ministry of Economic Affairs 2010). In the NREAP, the Dutch government highlights wind and biomass co-firing in power plants as areas of high priority, because of their affordability (Ministry of Economic Affairs 2010). As noted, however, the SDE-plus scheme does not provide subsidies for biomass co-firing projects. Subsidies for biomass co-firing do not stimulate technological innovation or the transition to a low-carbon economy: on the contrary, such subsidies could extend the use of coal-fired power plants and provide undesired technological lock-in effects. Several plants still receive subsidies from long-term contracts, but some plants that no longer receive subsidies have stopped co-firing.

The parties to the 2013 Energy Agreement agreed that promoting the use of biomass co-firing in coal-fired power stations was not to exceed the level of 25 petajoule, a target linked to the planned shutdown of five coal-based power plants built in the 1980s (three to be closed down with effect from 1 January 2016, and two more as of 1 July 2017). Biomass co-firing in coal-fired power plants might become less important as a source of renewable energy in the Netherlands, but the scope for expanding the share of liquid and solid biofuels will depend on the outcome of the EU agreement on sustainability criteria for biomass. The IEA maintains that co-firing of biomass, or full biomass conversion of inefficient coal plants, can provide innovative alternatives to shutting down existing Dutch coal plants (IEA 2014, p. 168).

The Netherlands was among the first leaders with demonstration windpower facilities in the 1990s, but developments in neighbouring countries, particularly Denmark and Germany, have since outpaced the Netherlands. On-shore wind is one of the cheaper renewable options, but local opposition and long planning processes are obstacles to rapid, large-scale realization. In 2013, the Netherlands had a total of 2434 MW of installed windpower capacity, of which 2206 MW was on-shore (IEA 2014, pp. 109–10). The target in the Energy Agreement is 6000 MW installed power capacity from on-shore wind turbines by 2020.

Off-shore windpower can be realized more quickly and with greater societal acceptance, but is a far more expensive option (Rathmann et al. 2010, p. 61). In 2012, two off-shore wind-farms were in operation, with a total of 228 MW installed capacity (IEA 2014, p. 110). As of June 2014, existing and planned off-shore windpower capacity amounted to a total of around 1000 MW. The Energy Agreement target is 4450 MW installed power capacity from off-shore wind turbines by 2023 (Energy Agreement 2013).

Other sources of renewable energy are solar and geothermal energy, but according to the Dutch government these do not hold the same potential as bio-energy generated from organic material (biomass), and on-shore as well as off-shore windpower. The NREAP also includes several other measures, such as tax relief, grid priority and more effective permit-granting procedures.

The Dutch government has claimed that the Netherlands is on track to meet its 14 per cent renewable energy target by 2020 (Ministry of Economic Affairs 2013, p. 37), but the country is lagging behind its interim goals for 2011 and 2012 (European Environment Agency 2014). In addition to the SDE-plus scheme, the government has earmarked a budget for renewable energy spending for the period up to 2020, set to rise from €900 million in 2013 to €3.8 billion by 2020. Given its increased spending on renewable energy, the Dutch government maintains that the more ambitious national target of 16 per cent renewable energy in 2023 is feasible (Ministry of Economic Affairs 2013, p. 37). However, the Netherlands is struggling to increase the share of renewable energy sources (European Environment Agency 2014).

The share of renewables in final energy consumption in the Netherlands was only 4.5 per cent in 2013 (Hekkenberg and Verdonk 2015). Biomass and waste accounted for more than 70 per cent of all renewable energy, and windpower for almost 20 per cent (Energy Agreement 2013, p. 6). As biomass is used primarily for co-firing in coal-fired power plants in the Netherlands, the country has a long way to go to attain its declared goal of taking a big step in the transition towards a more sustainable energy system by 2020. Hydropower, solar energy, geothermal energy and others make only a small contribution to the share of renewables in final energy consumption.

As argued in a report commissioned by the Green European Foundation, an environmental NGO based in Brussels, 'its once strong position in the areas of RES research and product development is weakening, and the chances for Dutch entrepreneurs to obtain leading market positions are declining' (Rathmann et al. 2010, p. 61). With existing measures, the renewables share in 2020 is expected to be 10.6 per cent. Including additional measures, the share in 2020 is expected to be 12.4 per cent, within a range of 10.5 to 13 per cent (Hekkenberg and Verdonk 2015) – still below the 14 per cent target by 2020 set by the RED.

To summarize, the Dutch renewables sector has developed rather slowly. It is based primarily on biomass and waste, although windpower is projected to increase towards 2020. Renewable energy subsidies (SDE and SDE-plus), regulation of renewable energy in transport, energy performance standards for buildings and tax measures are expected to speed up the development of this sector, but the Netherlands is lagging considerably behind its EU targets as well as in the pace of renewables deployment in neighbouring countries.

CCS Directive

The Netherlands has transposed the CCS Directive fully, with only a slight delay in the timetable for implementation. By September 2011, all the necessary legal documents had been amended and entered into force. The first CO_2-storage licence (the P18-A field in the North Sea) in accordance with the new legislation was issued in July 2013. This was the first storage licence in the EU subjected to the advisory procedure by the Commission (on the basis of the CCS Directive).

However, Dutch CCS ambitions have been reduced because of low public acceptance and funding for demonstration projects, as well as project delays. CCS is still factored in as part of ESD implementation in the Netherlands but, given the lack of CCS progress, including unrealized CCS projects in ESD, such calculations seem questionable. The Netherlands has sought to set up CCS demonstration projects, but these have suffered various setbacks. In 2010, one big CO_2 transportation and storage demonstration project planned by Shell in the densely populated area of Barendrecht was cancelled due to massive local opposition and delays in obtaining permits (see Feenstra et al. 2010; van Noorden 2010).

The intention had been to send CO_2 emissions through a pipeline from a gasification hydrogen plant at Shell's Pernis refinery near Rotterdam, about 20 km away. After heavy local opposition to the Barendrecht project, the government decided not to set up land-based projects for carbon capture, use and storage projects in any form, but to concentrate solely on off-shore CCS (Notenboom et al. 2012, pp. 17–18).

Preparations for an off-shore CCS demonstration project near the Rotterdam harbour area (the ROAD project) are ongoing, and efforts focus on achieving robust funding for the project. The project has received €180 million from the EU and €150 million from the Dutch government, but actual deployment has encountered significant difficulties due to the high costs involved. No other demonstration projects are planned, and no projects have been funded by NER 300, the EU's main funding programme for CCS and innovative renewable energy projects.

CCS projects do not receive much attention in the 2050 roadmaps of the Netherlands (Notenboom et al. 2012). In the national Energy Agreement (2013), however, it was agreed that CCS would be necessary to achieve an entirely sustainable energy-supply system, and the government has pledged to take steps towards producing a long-term strategy on the role of CCS in the transition to such an energy-supply system (Energy Agreement 2013, p. 10).

Conclusions

The Netherlands has been relatively efficient in transposing the various parts of the EU package. As regards *application* of the EU package and progress towards the 2020 targets, there has been correspondence between the EU 2020 targets and policy/behavioural change, but the Netherlands is still struggling to increase the share of renewable energy and apply CCS.

When assessed with a view to long-term decarbonization, investments in fossil-power generation overshadow investments in renewable energy. As of 2013, the Netherlands had eight coal-fired power plants, built between 1980 and 1994, with combined capacity of 4.2 gigawatts (GW) – representing 15.5 per cent of total installed capacity (IEA 2014, p. 165). Co-firing of coal and biomass has been the most important renewables measure. Although five coal-based power plants built in the 1980s will be shut down, the construction of three new plants with combined capacity of 3.5 GW (IEA 2014, p. 165) will lock in future carbon emissions. According to the IEA, however, 'the investment climate in the Netherlands has recently changed, putting at risk any new investment in clean and efficient coal-fired power generation' (IEA 2014,

p. 165). This may mean that co-firing of coal and biomass will not be attractive as a renewables measure in the long term.

As noted, the Netherlands is one of the EU's most fossil-fuel-dependent member-states, with over 90 per cent of its energy mix based on fossil fuels. In April 2014, the IEA urged the Netherlands to increase its deployment of renewable energy and to develop the last of its gas fields as well as its shale-gas resources (ENDS 2014). Although there are plans for increasing the renewables share by 2020, we can safely conclude that the EU climate and energy package has not yet been a game-changer in the Netherlands. On the other hand, the Climate Agenda for 2030 reaffirmed the commitment in the Energy Agreement to achieve CO_2 emissions reductions of between 80 per cent and 95 per cent by 2050 compared with 1990, as well as the need for more forceful action on climate mitigation and adaptation (Ministry of Infrastructure and the Environment 2013).

8.5 EXPLAINING IMPLEMENTATION

Fit and Adaptation Pressure

The Netherlands is one of the few EU member-states to have supported the EU ETS since its inception. Emissions-trading has remained a key priority for all Dutch government coalitions, with backing from industry. With the help of Germany, Dutch concerns for energy-intensive industries were taken into account in the negotiations, as almost all energy-intensive industry sectors were exempted from auctioning. Like several other EU countries, the Netherlands has a strong interest in maintaining a level international playing field for its energy-intensive industries. It has had various climate policies in place for dealing with emissions from non-ETS sectors, and the national 16 per cent reduction target by 2020 was accepted during the negotiations. The Netherlands is on track as regards meeting this target.

The RED was not particularly favoured by the Netherlands during the 2008 negotiations, but the Dutch renewable energy subsidy scheme (SDE and SDE-plus) was in line with EU renewables requirements. The Dutch 14 per cent renewables target, although well below the national target of 20 per cent by 2020 adopted in 2007, was ambitious compared with the national baseline situation with heavy dominance of fossil fuels in the energy mix. Accordingly, the 'fit' was mixed, and adaptation pressure rather high. On the other hand, the Netherlands has experienced more significant implementation challenges than could be expected from the 'degree of misfit' approach.

EU CCS policies were in line with Dutch policies and the negotiating position. CCS would be particularly suitable for coal-fired power plants and for the country's large refining sector – but, as explained above, the Netherlands has cancelled its CCS pilot project.

We expected that parallel negotiations of different EU legislation would lead to variation in implementation as a result of trade-offs between different national priorities in the negotiations. There has clearly been variation in implementation, but we find no evidence to indicate that this is due to a balancing of different priorities. What we know is that the Netherlands was not enthusiastic about the renewables target, but it did not demand any significant changes to the Commission proposal, except for more generous free allocation rules for energy-intensive industry.

With regard to the EU package and the 20–20–20 targets as a whole, the EU sought to combine policy solutions for different concerns in a package approach, including climate change, energy affordability, security of supply and international competitiveness. The Dutch energy-economic situation and policy priorities partly matched the situation and priorities in the EU. At the time the EU package was adopted, concerns about climate change were rather high on the political agenda in the Netherlands, as were concerns regarding energy affordability and international competitiveness.

Unlike the situation in several other EU countries, security of supply has not been a major concern in the Netherlands, but this may be about to change. For many years, the Netherlands has been a net exporter of gas from its huge Groningen gas field. However, the country is expected to shift from being a net exporter to a net importer of gas by 2025, due to declining production from the Groningen field. The IEA has recommended that the Netherlands develop the last of its gas fields, as well as its shale-gas reserves, in the coming years, but strong public and local government opposition may obstruct this (ENDS 2014). Exploration and test drilling for shale gas have been put on hold, pending the outcome of a national impact assessment.

In sum, we find that the 'fit' between national and EU policies and the correspondence between the Dutch negotiating position and the revised ETS Directive were high, resulting in low adaptation pressure. Here our main conclusion is that 'misfit' and adaptation pressure between the EU package and the situation in the Netherlands cannot explain the rather slow implementation of parts of the package, the RED in particular. We must delve more deeply into the role of domestic politics in order to explain the implementation challenges in the Netherlands.

Domestic Politics

Governmental actors

The Netherlands is a parliamentary democracy. The head of government is the prime minister, who usually is the leader of the largest party in the governing coalition. The government is responsible to the bicameral parliament, which has legislative powers. Typical of Dutch politics and governance is the focus on achieving broad consensus on important issues, within the political community and in society.

Can changes in (majority) government explain changes in political priorities with consequences for implementation? The short answer is 'yes'. Since 2008, the Netherlands has implemented EU climate and energy policies under the *Clean and Efficient* programme (adopted in 2007) and various energy transition policies, including SDE-plus. The country's policies were based on the EU climate and energy package and the EU 20–20–20 targets, but the centre-left government (the fourth Balkenende government) had established more ambitious targets than required for the Netherlands by the EU. However, the Dutch centre-right coalition minority government that took office in 2010 (the first Rutte government) lowered the targets for the level of GHG emissions and the share of renewable energy to the level required by the EU. The centre-right government also distanced itself from the more ambitious GHG targets that had been established by the centre-left government in 2007. A key difference between the 30 per cent target established by the fourth Balkenende government and the 16 per cent target established for the Netherlands by the ESD is that while the former applies to all GHG emissions, the latter applies only to non-ETS emissions. Changing governments with different priorities have also represented a problem for increasing the renewables share. Indeed, instability in the climate and energy policies of shifting government coalitions has been a recurrent challenge in the Netherlands.

On the whole, Dutch implementation ambition levels have varied with changes in government. Instability in implementation priorities can also stem from differing positions within the government. Two separate ministries are in charge of climate policy and energy policy: the Ministry of Infrastructure and the Environment is responsible for climate policy, and the Ministry of Economic Affairs for energy policy. This division of responsibilities mirrors the situation in many other countries, and helps to explain why it has been hard to integrate climate and energy policies in the Netherlands. The EU package approach does not seem to have affected the Dutch implementation process, because the package components have largely been implemented separately.

Affected societal actors

The Netherlands is a major producer of natural gas and has a major petroleum-converting industry based on oil refining and petrochemical products. Additional energy-intensive industries include other chemical, steel and metallurgic industries. These industries could be heavily affected, and negatively so, by a development towards a low-carbon economy based on carbon-pricing and renewables subsidies. Shell, as the biggest and most important company in the Dutch petroleum sector, appears to be the single most important actor in lobbying both EU and Dutch climate and energy policies. According to the EU's transparency register, Shell is the sixth biggest lobbyist in Brussels, spending between €4.25 and €4.5 million a year lobbying EU institutions (EU Transparency Register 2015). As we shall see, Shell lobbied to undermine EU renewable energy targets prior to the agreement on the 2030 climate and energy package. Shell's argument has been 'gas is good for Europe, and Europe is good at gas' – that is, a market-led strategy of gas expansion will help Europe in transitioning to a low-carbon energy system. CCS could have provided a major opportunity for the fossil-fuel industry to reduce emissions, but CCS development has suffered from stagnation. Costs are concentrated to the key economic sectors, but there is no substantial 'green' technology industry in the Netherlands.

The oil-refining industry is part of the large energy-intensive sectors in the Netherlands exposed to international competition. These industries are concerned about the indirect costs of putting a price on carbon, although they generally support emissions-trading. The Netherlands thus has a strong interest in maintaining a level international playing field for its energy-intensive industries, and is among the six EU member-states allowed to issue compensation for indirect carbon costs. The Dutch agenda of securing the interests of energy-intensive industries exposed to international competition is promoted by the Confederation of Netherlands Industry and Employers (VNO-NCW).

Dutch electricity producers, mostly owned by German companies, have seen a sharp decrease in the utilization of their fossil-fuel plants because of the surge in renewable capacity in Germany. The close connection between the German and Dutch electricity markets has resulted in oversupply of cheap electricity to the Dutch market and lower profits for Dutch electricity producers. Several recently built Dutch gas-fired power plants are not going to be put into operation because of the oversupply of electricity and reduced profits.

The dominant role of fossil fuels and energy-intensive industries appears to have limited the search for new low-carbon opportunities. The Netherlands was once a frontrunner in development and deployment of

CHP technology and CCS research and development, and could also have been a frontrunner in the development of on-shore and off-shore windpower, but it has lost its early lead to other countries. As mentioned above, it has been observed that the chances of Dutch clean-tech entrepreneurs to obtain leading market positions have declined. Yet, the parties to the 2013 Energy Agreement have aimed to place the Netherlands among the top 10 in credible clean-tech rankings by 2030. An intermediary aim, on the way to achieving the 2030 clean-tech target, is to quadruple the economic value of the clean-energy technology chain by 2020 compared with 2010. As regards employment, the aim is to create at least 90,000 additional fulltime equivalents between 2014 and 2020.

To summarize, the components of the package have been implemented mainly as separate policies in the Netherlands, with the 'package approach' apparently making little difference to domestic implementation. The implementation of climate *and* energy policies has always spurred domestic conflicts among societal interest groups in the Netherlands, as the two policy fields have partly conflicting goals and have been challenging to combine, and the EU package approach does not seem to have facilitated policy integration. On the other hand, internal effort-sharing between the ETS and non-ETS sectors represents a concrete example of cost-sharing. Implementation of the ESD rests on 'skilful accounting measures' as well, which has meant that some Dutch non-ETS sectors could be re-classified as ETS sectors, easing the pressure on these sectors.[3] National decision-makers do not seem to have fully exploited the potential synergies of the EU package – but neither has the package created difficulties for implementing existing national agreements and policy packages.

Policy style

A consensual policy style may improve the likelihood of successful implementation, because the target groups that are included in the policy process are more likely to implement agreed policies than those that are not. The Netherlands traditionally employs a consensual policy style. Lijphart (1977) has described the Netherlands as a 'consociational' democracy: a democratic form held to explain political stability in culturally segmented societies. A consociational democracy is characterized by the development of institutions that decrease the potential for conflicts, and the relationship between interest groups and regulators is one of cooperation and bargaining. The Dutch political system has clear neo-corporatist characteristics: the key to stability is cooperation and compromises with industry and other interest groups during policy-making and implementation. Strong and persistent cooperation between the state and industry is sometimes seen as an advantage as regards the development of environmentally friendly technology and implementation of policies. Due to the dominant role of fossil fuels and energy-intensive industries in the Netherlands, an inclusive and consensual approach means that the actors most negatively affected by an ambitious EU climate policy also have most influence in decision-making.

The consensual Dutch policy style is visible in the making and implementation of domestic climate and energy policy. One particular feature of the policy is the extensive use of covenants and agreements, as seen in the implementation of the ESD. Another feature is the extensive formal and informal inclusion of industry and interest groups in the making and implementation of Dutch climate and energy policy. The Dutch '*poldermodel*' – the consensus-based tripartite collaboration between employers' organizations, labour unions and the government in economic and social policymaking – was applied for the first time to the climate and energy field when the 2013 Energy Agreement was developed by the SER. As the SER has traditionally advised the Dutch government and parliament on social and economic legislation and policy, in deciding to act as a platform for the negotiations on the Energy Agreement, it entered into an entirely new field of policymaking and facilitation.

More than 40 societal, economic and environmental organizations took part in the negotiations, which also included the government. After more than six months of intense deliberations, the parties agreed on the comprehensive Energy Agreement, in which the Dutch state cooperates with a broad range of stakeholders to reduce emissions from all sectors. This arrangement has been held to be unique in terms of its comprehensiveness and inclusiveness and its holistic approach to climate and energy issues. The parties hope that the Energy Agreement will lead to greater stability and predictability for Dutch climate and energy policies, but this remains to be seen. In its assessment of Dutch energy policies, the IEA notes that it will be challenging 'to ensure the actual delivery of the many actions and commitments' in the Energy Agreement, and that implementation of actions might be delayed or fail if one party steps out of the agreement (IEA 2014, p. 28). As regards policy style, however, we can clearly see the Dutch preference for negotiated and comprehensive agreements and inclusive policymaking processes rather than traditional command-and-control regulations.

The question is, then, whether the EU package approach opened up for new alliances in Dutch policymaking processes, or cemented old ones. The making of the Energy Agreement was innovative in terms of the application of the *poldermodel* to the energy sector, even though it represents a continuation of a well-established and traditional policy style in the Netherlands. The EU package clearly influenced the broad scope of climate and energy issues, targets and the policy measures agreed by the many parties to the Energy Agreement. Previous more or less voluntary agreements have mainly dealt with specific issues and sectors. In addition, the EU package clearly increased domestic regulatory pressure and expectations that industry would adopt more progressive measures, which in turn motivated and convinced non-state actors to negotiate the Energy Agreement. Hence, the EU climate and energy package may be said to have been the decisive factor that motivated the establishment of the Energy Agreement. In a long-term perspective, the key societal groups that have to implement the EU package were also involved, through various voluntary agreements signed just before the package negotiations commenced in 2008.

The slow implementation of renewables policy in the Netherlands can be explained by dominant interests rather than privileged access. A consensual policy style (which we had expected would facilitate implementation) was not sufficient to deflect the strong counterforces that obstructed successful implementation. These forces were the energy-intensive industries and the energy producers' dependency on fossil fuels, which seem to have trumped the interests of the renewables industry in the Netherlands. That said – and contrary to expectations – it seems clear that the Dutch policy style helps to explain the rather slow development of renewables in the Netherlands. This consensual policy style seems to create societal acceptance for the implementation of government policies. In this sense, the consensual policy style is effective as a tool for creating legitimacy and acceptance of the government's climate and energy policy. However, because a consensual policy style makes it hard to challenge powerful actors such as the petroleum industry, it is probably less effective if the aim is to create radical policy change towards a low-carbon society. Our analysis indicates that the Dutch policy style serves to promote incremental, stepwise changes rather than large-scale industrial and societal transition.

Administrative organization

The relatively quick transposition process in the Netherlands appears to have been facilitated by a concentration of governmental responsibility. The EU Directives in the package were transposed at different times, but within a narrow timeframe. Only two ministries were involved in implementing all package components: the Ministry of Infrastructure and the Environment, and the Ministry of Economic Affairs. However, these ministries have somewhat differing interests and policy priorities. That could help to explain the mismatch of Dutch climate and energy policy and the lack of integration of the two policy fields.

The Ministry of Infrastructure and the Environment was created in 2010 following the merger of the former Ministry of Transport, Public Works and Water Management with the Ministry of Housing, Spatial Planning and the Environment. This merger could be seen as a degradation of the environmental policy field, since the portfolio of the new minister did not include environmental issues. Climate policy and other environmental issues were now to be handled by a new subordinated state secretary (a type of junior minister). These institutional changes reflected the policy priorities of the centre-right government that took office in 2010.

The Netherlands has three levels of administration: state, provinces (12) and municipalities (393). Local opposition to plans from the central government helps to explain the slow implementation. If central government policies are frequently blocked, this could cause vertical disintegration of climate and energy policies. Although the evidence does not indicate a general pattern of vertical disintegration in the Netherlands, local councils and communities have obstructed plans from the central government on several occasions. As noted, Royal Dutch Shell's CCS project in the Netherlands (Barendrecht) was cancelled by the government in November 2010 because of strong local protests and delays in obtaining permits. Residents and town officials opposed the plan because of safety concerns and what they regarded as an experimental project with potentially harmful consequences for the city and local community. After various setbacks with CCS demonstration projects, the Dutch government decided not to pursue CO_2 storage on land in any form, but to concentrate on off-shore CCS only.

Local opposition and long planning processes also represent problems for rapid and large-scale realization of on-shore windpower, which is one of the cheaper renewable options. Large wind-farms are deemed superior to stand-alone windmills scattered over extensive areas. The Netherlands is a small and densely populated country, which makes it difficult to find suitable locations for wind-farms. The exception is the sparsely populated province of Flevoland, which is full of windmills.

On the whole, horizontal and vertical distribution of competences helps to explain the slow implementation of parts of the EU package in the Netherlands, particularly the CCS and RED Directives.

8.6 CONSEQUENCES FOR NATIONAL POSITIONS TOWARDS 2030

Resistance to new policies can be expected if the burden- and opportunity-sharing that can promote EU-level agreement is not reproduced domestically. Conversely, support can be expected if member-states experience positive consequences from implementation.

In March 2013, the Commission presented its Green Paper on a 2030 framework for climate and energy policies, and asked for insights and viewpoints through a public consultation process (Commission 2013). Dutch interests were deeply divided over this framework. The PBL Netherlands Environmental Assessment Agency argued for the continued mix of complementary 2030 targets on GHG emissions reduction, renewable energy and energy efficiency, as well as specific targets for innovative low-carbon technologies (Koelemeijer et al. 2013). The parliamentary group of the liberal party Volkspartij voor Vrijheid en Democratie (VVD) – the largest party in the Lower House of the Dutch parliament and part of the government – argued forcefully that the only target needed for 2030 was a single CO_2 emissions target (VVD 2013). This was supported by the Dutch Ministry of Economic Affairs and the VNO-NCW.

The preference for a single GHG target has been combined with support for the EU ETS as the main instrument. This is largely due to free allowances for energy-intensive industry and the generous import of external credits aided by a very low carbon price. In its Communication to the meeting of the Council (Environment) on 12 June 2014, the Netherlands confirmed its strong support for the EU ETS as 'the most important building block of climate policy', while adding that 'discontinuation of the CO_2 reduction target for transport fuels in the Fuel Quality Directive is a move in the wrong direction and should be corrected' (European Council 2014, p. 4).

Why did continuation of the package approach not attract wider support? The reason appears to be linked to the lack of domestic cost-sharing and failure on the part of domestic industry to exploit new opportunities. Dutch renewables policy has not yet created any significant new businesses, and CCS has been virtually abandoned, at least in the short term.

The Dutch Environmental Minister joined Environmental Ministers from other EU countries in the Green Growth Group in urging the European Council in March 2014 to agree on the core elements of the 2030 framework as set out by the Commission (Green Growth Group

2014). However, the Netherlands teamed up with the UK and other member-states that resisted binding targets for individual member-states on energy efficiency and renewable energy. These targets did not make it into the final agreement on EU climate and energy policies for 2030, agreed in October 2014. Instead, the EU agreed on a rather weak target of at least 27 per cent renewable energy, binding only on the EU as a whole.

In April 2015, *The Guardian* revealed that Shell had lobbied to undermine EU renewable energy targets since 2011. In a five-page letter to Commission President Barroso in October 2011, Shell had maintained that a market-led strategy of gas expansion would save Europe €500 billion in its transition to a low-carbon energy system, compared with an approach centred on renewables (*The Guardian* 2015). Dutch opposition against renewable energy targets that would be binding for individual member-states can be understood primarily by the climate policies of the Rutte centre-right government. These policies have likely been influenced by pressure from the fossil-fuel industry and the problems experienced in significantly increasing the renewables share in the Netherlands, as required by the EU RED.

The government's rather unambitious climate policies made the international headlines in June 2015, when The Hague District Court ruled that the government must take more action to reduce GHG emissions in the Netherlands. The government has to ensure that Dutch emissions in 2020 will be at least 25 per cent lower than those in 1990. With the government's current policy, the Netherlands in on track for a 17 per cent reduction at most in 2020 and it lacks plans for a further cut, according to the ruling. This landmark verdict is the first time a government has been ordered to raise its climate ambition by a court (Rechtbank den Haag 2015). On 1 September 2015, the Dutch government announced that it would appeal the ruling, but at the same time said it would start taking measures to comply with the ruling (ENDS 2015).

8.7 CONCLUSIONS

The Netherlands has had climate and energy policies in place for many years, but has been struggling to integrate the two policy fields. With new and more ambitious climate policies adopted in 2007, the country appeared well prepared for the EU climate and energy package. Targets and measures in the package were essentially similar to domestic targets and measures, so the EU package was largely compatible with existing Dutch climate and energy policies. The Netherlands was exposed to relatively low adaptation pressure from the EU, although concerns had been raised at the domestic level with regard to the lack of progress on renewables.

We also find that the EU climate and energy package has provided greater stability for Dutch climate and energy policies, establishing both short-term and longer-term targets. The lack of stability and long-term vision in Dutch climate and energy policies has often been cited as a reason for the slow development of renewable energy and new technologies in the Netherlands. The EU package has reduced the discretion of member-states, although some room for manoeuvre still exists. Without the EU package, Dutch climate and renewables policies would have ranked rather low on the political agendas of recent government coalitions. Under shifting government coalitions with changing priorities, EU climate and energy policies have provided some stability and certainty for Dutch policies. Indeed, the EU 2020 package appears to have been crucial for avoiding further reduction of climate ambitions and renewables targets, since the Dutch centre-right government that took office in 2010 lowered the national targets for the level of GHG emissions and the share of renewable energy to the levels required by the EU. Similarly, the EU's 2030 policy framework for climate and energy is necessary for supporting low-carbon and renewables investments in the Netherlands.

How, then, can we explain the fact that the Netherlands appears to be struggling to achieve the target in the Renewables Directive and the transformation to a low-carbon society? The energy-economic situation and policy priorities in the Netherlands were largely compatible with the situation and priorities in the EU as a whole – indicating that 'misfit' in energy-economic situations and policy priorities cannot explain implementation challenges.

The oil and gas industry and the fossil-fuel-dependent petrochemical industry as well as other energy-intensive industries have strong interests in maintaining energy affordability, and can thus be regarded as industry interests that may block a transition to a low-carbon economy. Although their interests in maintaining the traditional carbon economy go some way in explaining the slow development of the renewables sector in the Netherlands, various Dutch governments have never proposed radical low-carbon solutions. Factors like population density, topography and lack of high hydropower or geothermal potential also help to explain the slow development of the sector.

We have seen how strong popular and local government opposition has effectively blocked large wind-farms and on-shore CCS demonstration projects, as well as shale-gas exploration and test drilling in the case of fossil fuels. Local governments and local communities could thus be regarded as veto players that have obstructed the implementation of certain national energy policies, in the case of both renewables and fossil fuels.

Regarding the Dutch policy style, we have noted how the Netherlands is referred to as a consociational democracy with strong neo-corporatist characteristics: the key to stability is cooperation and compromises with industry and other interest groups during policymaking and implementation. Accordingly, the use of covenants – negotiated agreements with industry – is a typical feature of Dutch climate and energy industry. The Dutch policy style promotes small, incremental changes rather than large-scale industrial transition – and this appears to have resulted in rather slow implementation of the package components that required substantial Dutch policy changes, particularly as to renewables.

Finally, with respect to administrative organization, we observe that only two ministries (infrastructure/the environment, and economic affairs) were involved in implementing all package components. While the relatively quick transposition process appears to have been facilitated by concentration of governmental responsibility, the Ministry of Infrastructure and the Environment and the Ministry of Economic Affairs have somewhat differing interests and policy priorities. This might help to explain the lack of integration of climate and energy policies in the Netherlands.

The EU package approach does not seem to have facilitated the integration of domestic climate and energy policies. On the other hand, it has not disrupted existing policies and agreements in any major way. The directives and decisions in the package have essentially been transposed and implemented in separate processes in the Netherlands. We may thus conclude that while each EU directive, decision and regulation has had a significant impact on Dutch climate and energy policies, the fact that these all came as an EU 'package' did not make a significant difference.

The Netherlands has lost its early lead in climate and renewable policies. Beyond its good transposition score, it has been rather slow in implementing parts of the EU climate and energy package, the RED in particular. Lack of stability and fragmentation of responsibilities have characterized Dutch climate policies. This may be about to change, however, as climate and renewable energy policies have reappeared high on the political agenda. With the 2013 Energy Agreement, broad societal consensus on the need for stability and long-term vision has been manifested in a comprehensive climate and energy document. What remains to be seen is whether the parties to the agreement will prove willing and able to deliver on their promises.

NOTES

1. Prime Minister Jan Peter Balkenende had headed three earlier cabinets that could better be described as centre-right.
2. Coalition agreement *Building Bridges* (29 October 2012) of the Rutte–Asscher government.
3. Netherlands Article 24 application in accordance with authorization of additional allowances under Article 24 (2) of the EU ETS Directive.

REFERENCES

Andersen, M.S. and D. Liefferink (1997), 'Introduction: the impact of the pioneers on EU environmental policy', in M.K. Andersen and D. Liefferink (eds), *European Environmental Policy: The Pioneers*, Manchester: Manchester University Press, pp. 1–39.

Commission (2013), Green Paper: *A 2030 Framework for Climate and Energy Policies*, 27.3.2013, COM (2013) 169 final, Brussels: European Commission.

Commission (2014), *Progress Towards Achieving the Kyoto and EU 2020 Objectives (October)*, Brussels: European Commission.

Ecofys (2015), *History of Carbon Capture and Storage in the Netherlands*, accessed 4 June 2015 at http://ccs-roadmap.ecofys.com/index.php/History_of_CCS_in_the_Netherlands.

ENDS (2008a), 'EU states urge more flexibility in climate plans', 3 March.

ENDS (2008b), 'Carbon leakage concerns dominate EU ETS debate', 11 September.

ENDS (2008c), 'France accused of "attacking" EU renewables goal', 3 October.

ENDS (2008d), 'Still no consensus in sight on EU climate plans', 21 October.

ENDS (2008e), 'Dutch agency issues warning over climate goals', 7 November.

ENDS (2014), 'Netherlands falling behind on renewables – IEA', 24 April.

ENDS (2015), 'Dutch government to appeal landmark climate ruling', 2 September.

Energy Agreement (2013), *English Summary, Energieakkord voor duurzame groei*, The Hague: Social and Economic Council of the Netherlands (SER), accessed 12 January 2015 at http://www.energieakkoordser.nl/energieakkoord.aspx.

EU Transparency Register (2015), 'Shell companies', accessed 5 May 2015 at http://ec.europa.eu/transparencyregister/public/consultation/displaylobbyist.do?id=05032108616-26&isListLobbyistView=true.

European Council (2008), *Proposal for a Directive of the European Parliament and of the Council on the geological storage of carbon dioxide and amending Council Directives 85/337/EEC, 96/61/EC, Directives 2000/60/EC, 2001/80/EC, 2004/35/EC, 2006/12/EC and Regulation (EC) No 1013/2006*, Brussels: General Secretariat of the Council.

European Council (2014), *Communication from the Commission on A Policy Framework for Climate and Energy in the Period from 2020 to 2030. Replies Received from the Czech Republic and the Netherlands to the Presidency Questions*, 11 June, Brussels: General Secretariat of the Council.

European Environment Agency (2013), *Trends and Projections in Europe 2013: Tracking Progress towards Europe's Climate and Energy Targets until 2020*, EEA Report No. 10/2013, Copenhagen: European Environment Agency.

European Environment Agency (2014), *Trends and Projections in Europe 2014: Tracking Progress towards Europe's Climate and Energy Targets for 2020*, EEA Report No. 6/2014, Copenhagen: European Environment Agency.

Faber, J., M.J. Blom, S.M de Bruyn, D. Nelissen, S.J. Aarnink, A. de Buck, D. Bennik, F.H. Oosterhuis and O.J. Kuik (2012), *Het Nederlandse klimaat- en energiebeleid: analyse van beleidsevaluaties 1989–2012*, Delft: CE Delft.

Feenstra, C.F.J., T. Mikunda and S. Brunsting (2010), *What Happened in Barendrecht? Case Study on the Planned Onshore Carbon Dioxide Storage in Barendrecht, the Netherlands*, Amsterdam: Energy Research Centre of the Netherlands.

Green Growth Group (2014), Green Growth Group Ministers' statement on climate and energy framework for 2030, joint statement on 3 March 2014, accessed 4 February 2015 at https://www.gov.uk/government/news/green-growth-group-ministers-statement-on-climate-and-energy-framework-for-2030.

Hekkenberg, M. and M. Verdonk (2015), *Netherlands: National Energy Outlook 2014, Summary*, February, ECN, PBL, Statistics Netherlands, Netherlands Enterprise Agency, accessed 4 May 2015 at http://www.pbl.nl/en/publications/netherlands-national-energy-outlook-2014.

Howes, T. (2010), 'The EU's Renewable Energy Directive', in S. Oberthür and M. Pallermarts (eds), *The New Climate Policies of the European Union*, Brussels: VUB Press, pp. 117–50.

IEA (2014), *Energy Policies of IEA Countries – The Netherlands 2014 Review*, Paris: OECD/IEA.

Koelemeijer, R., J. Ros, J. Notenboom and P. Boot (2013), 'Comments on the EC Green Paper *A 2030 Framework for Climate and Energy Policies*', PBL Publication No. 1158, Bilthoven/The Hague: PBL Netherlands Environmental Assessment Agency.

Liefferink, D. (1997), 'The Netherlands: a net exporter of environmental policy concepts', in M.K. Andersen and D. Liefferink (eds), *European Environmental Policy: The Pioneers*, Manchester: Manchester University Press, pp. 210–50.

Lijphart, A. (1977), *Democracy in Plural Societies: A Comparative Exploration*, New Haven, CT: Yale University Press.

Ministry of Economic Affairs (2010), *National Renewable Energy Action Plan, Directive 2009/28/EC*, The Hague: Ministry of Economic Affairs.

Ministry of Economic Affairs (2013), *Nederlands Nationaal Hervormings-programma 2013*, The Hague: Ministry of Economic Affairs.

Ministry of Infrastructure and the Environment (2011), *Strategy on Climate Policy En Route to 2020* (Kabinetsaanpak Klimaatbeleid op weg naar 2020), The Hague: Ministry of Infrastructure and the Environment, accessed 15 January 2016 at http://www.rijksoverheid.nl/documenten-en-publicaties/kamerstukken/2011/06/08/kabinetsaanpak-klimaatbeleid-op-weg-naar-2020.html.

Ministry of Infrastructure and the Environment (2013), *Climate Agenda for 2030*, The Hague: Ministry of Infrastructure and the Environment, accessed 4 February 2015 at http://www.government.nl/news/2013/10/04/climate-agenda-mitigation-adaptation-and-business-sense.html.

Noorden van, R. (2010), 'Buried trouble', *Nature*, 463 (18), 871–3.
Notenboom, J., P. Boot, R. Koelemeijer and J. Ros (2012), *Climate and Energy Roadmaps towards 2050 in North-Western Europe*, Bilthoven/The Hague: PBL Netherlands Environmental Assessment Agency.
Peeters, M. (2011), 'Climate law in the Netherlands: an overview of current mitigation measures', *Ympäristöjuridiikka*, 3 (11), 125–34.
Rathmann, M., T. Winkel and R. de Vos (2010), 'Evaluation of the Dutch NREAP', in Green European Foundation (ed.), *27 National Action Plans = 1 European Energy Policy? An Analysis of Six Renewable Energy Action Plans*, Brussels: The Green European Foundation, pp. 56–64.
Rechtbank Deen Haag (2015), ECLI:NL:RBDHA:2015:7196, accessed 27 August 2015 at http://uitspraken.rechtspraak.nl/inziendocument?id=ECLI:NL:RBDHA:2015:7196.
Skjærseth, J.B. and J. Wettestad (2008), *EU Emissions Trading: Initiation, Decision-Making and Implementation,* Aldershot: Ashgate.
The Guardian (2015), 'Shell lobbied to undermine EU renewable targets, documents reveal', accessed 7 May 2015 at http://www.theguardian.com/environment/2015/apr/27/shell-lobbied-to-undermine-eu-renewables-targets-documents-reveal.
VVD (2013), 'Contribution to the Green Paper "A 2030 framework for climate and energy policies"', The Hague: VVD, accessed 26 March 2015 at http://ec.europa.eu/energy/consultations/20130702_green_paper_2030_en.htm.

9. Implementation in Norway

Torbjørg Jevnaker

9.1 INTRODUCTION

Cooperation between Norway and the European Union has a long history, also as regards climate and energy issues. Norway is not an EU member-state, but has been part of the Single Market through the Agreement on the European Economic Area (EEA Agreement) since 1994. The prerequisite for market access is implementation of EEA-relevant EU legislation. The EU's climate and energy package for 2020 fell into that category. Norway's lack of full EU membership meant that it could not participate in the 2008 negotiations on the climate and energy package. During negotiations within the European Council, some member-states were able to secure concessions crucial for their accepting the package (see Chapter 5). As a package catering to needs and interests at the EU level, a major aim was to reduce CO_2 emissions from fossil-fuel-dependent power and heating installations. In Norway, however, the power and heating sectors were already basically decarbonized – but emissions were growing, particularly within the petroleum sector. Given Norway's different situation and inadequate representation in EU discussions, it was to be expected that the package would face implementation hurdles.

Implementation of the EU climate and energy package in Norway has involved a mixed pattern of support, delays and opposition.[1] Moreover, there has been little behavioural change thus far. Thus, this chapter asks: How was the EU's climate and energy package implemented in Norway, and what can explain the differentiated implementation performance of various parts of the package? As a central aspect at the EU level was the linking of different policies into one package in order to achieve the 2020 targets with a view towards 2050, the implications of the package for Norway's long-term position will also be discussed here.

9.2 NORWAY'S CLIMATE AND ENERGY POLICIES UP TO 2007

The Norwegian economy has experienced relatively high economic growth since the 1980s, largely due to the petroleum sector, which is responsible for a quarter of overall value creation and a third of state revenues (Ministry of Petroleum and Energy (MPE) & Norwegian Petroleum Directorate (NPD) 2013, p. 21). Norwegian emissions rose from 50.4 million tonnes (mtoe) CO_2 equivalents in 1990 to 54.3 mtoe in 2005, with steepest growth within the petroleum sector (Statistics Norway 2013). By source, Norway's domestic energy consumption in 2005 was split between fossil fuels (58 per cent) and renewables (42 per cent) (Eurostat 2007, p. 459). About 99 per cent of on-shore electricity production came from renewable energy sources, predominantly hydro (Eurostat 2007, pp. 459–60). Some electricity is traded with EU member-states, but this is dwarfed by gas exports, which make Norway the second-largest gas supplier to the EU (Norwegian Oil Industry Association 2010).[2] In contrast to most other European countries and the EU in general, then, Norway is a net exporter of energy. A major consumer of electricity is the energy-intensive on-shore industry, relatively large in a European context, where Norway is the largest aluminium producer (Official Norwegian Report (NOU) 2012, p. 550).

Value creation and employment have been the overarching objectives of Norwegian energy policy. Moreover, energy policy has been seen in context with industrial policy, as the electricity sector and energy-intensive industries have historically developed in tandem (Wicken 2011). Policymakers have emphasized public ownership, strict regulations and high tax levels for the country's two major energy sectors (the petroleum sector and the electricity sector). This is in accordance with a strong emphasis on allocating the resource rent to the public (NOU 2012, pp. 548–50). As Norway has ample energy resources, security of supply has not been an issue beyond ensuring adequate transmission. In terms of renewable energy, an earlier windpower support scheme that had failed to deliver results triggered calls from the power producers for a new support scheme in order to incentivize investments in renewable energy (Boasson 2015). While the Ministry of Petroleum and Energy had sought to develop an alternative scheme (MPE 2006a, b), the scope for manoeuvre was restricted by state-aid guidelines from the EU. Wanting to avoid lengthy discussions with the EFTA Surveillance Authority (ESA), the Norwegian government decided to back a market-based certificates measure, entering into discussions with Sweden on establishing a joint market for green certificates (Boasson 2015). Agreement could not be reached on the overall target, and Sweden established a separate market in 2003. In late 2007, however, bilateral talks were resumed (MPE 2007).

Norwegian climate policy has placed a high emphasis on cost-efficiency, with strong political support for reducing emissions where this is cheapest and within an international framework. Thus, Norway has supported a global climate agreement as well as flexible mechanisms for offsetting (Ministry of the Environment (MoE) 2012b). Nevertheless, there has been recurrent debate on whether to mitigate climate change through domestic or international measures. National emissions have been targeted by general economic measures, primarily the CO_2 tax, emissions-trading, and fuel and car taxes. The CO_2 tax covered 55 per cent of Norwegian emissions, including the petroleum and transport sectors. Norway originally intended to set up an emissions-trading system for sectors not covered by the CO_2 tax (like steel, cement and refining), but in 2005 set up a system in readiness for the EU ETS instead. Norway has been part of the EU ETS since 2009 (NOU 2012, p. 98). In the transport sector, Norway has used the car tax to incentivize consumers to purchase less carbon-intensive automobiles, including electric vehicles (MoE 2007, p. 68). Finally, the country's traditional on-shore energy-intensive industries have been shielded from costly climate policy, although there were voluntary agreements for emissions reduction (Environment Agency 2014a).

At this crossroads between energy policy and climate policy, between growing emissions from important economic sectors like petroleum on the one hand and climate ambitions on the other hand, Norway has resorted to two main strategies. One has been the use of flexible mechanisms, which have allowed Norway to compensate its domestic growth in emissions through international offsetting. The other has been R&D programmes aimed at unlocking technological change that could resolve the tension between economic growth and increases in emissions. Particular emphasis has been given to CCS, which has been an important aspect of Norwegian climate policy as regards the petroleum sector (Tjernshaugen and Langhelle 2011, p. 180). In order to increase oil extraction, carbon has been stored on the Norwegian Continental Shelf since 1996 (MoE 2012b, p. 96); later, the government started working together with business to realize CCS on gas-fired power plants – which became necessary to make the power plants politically acceptable (Tjernshaugen and Langhelle 2011, p. 180).

9.3 NEGOTIATING THE PACKAGE: 2007–2008

Norway cooperates extensively with the EU through the EEA Agreement. Norway may participate as an observer in the expert committees and working groups of the European Commission (henceforth: Commission) (Ministry of Foreign Affairs (MFA) 2012, p. 18), but has no formal access to negotiations within the European Parliament or the Council – although Norwegian representatives may be invited to informal meetings on a case-by-case basis. The EEA Agreement does not give Norway access to participating in overarching strategic discussions with the EU, as there is no institutional framework for discussing cross-sectoral policymaking (NOU 2012, p. 860). Thus, Norway has greatest scope for influence through the EEA Agreement in the early stages of the individual legislative processes, notably in the initial policy-preparation phase managed by the Commission.

While Norway contributed to the Commission's policy preparation, the later stages in the EU's decision-making process on the climate and energy package posed several challenges. First, the linking of several pieces of legislation into a package increased the need for coordination between different parts of the Norwegian government. As a package encompassing the entire economy, ministries representing various sectors would need to exchange information and coordinate positions. Second, the EU process was unusual in the sense that it was fast-tracked, with only single readings in Council and the European Parliament – as well as with the heavy involvement of the European Council. With speedy and informal negotiations within and between the EU institutions, the potential for Norwegian influence was significantly reduced.[3]

For Norway, even keeping track of the EU negotiations, which took place in multiple fora and involved many actors and policy areas, proved challenging. In order to deal with the speedy and cross-sectoral policymaking at EU level, provisional steps were taken in Norway to facilitate exchange of information among the various ministries responsible for the affected policy-areas, including the Ministry of Petroleum and Energy, and the Ministry of the Environment. Beyond the already existing fora for interministerial coordination on EU issues, informants noted that telephone conferences on the climate and energy package were coordinated by the Ministry of Foreign Affairs. These meetings gathered representatives from multiple ministries as well as the Norwegian delegation in Brussels. However, one informant emphasized the lack of a strong organizational framework for cross-sectoral coordination on EU issues within the Norwegian executive, resulting in slow processes due, inter alia, to diverging views among different ministries.

The Norwegian government developed positions on the various elements of the package. It supported the revised ETS and the CCS Directives; while generally positive to the RED (Stortinget 2008b), Norway remained strongly opposed to binding national-level targets for renewable energy as part of that Directive (Europaportalen 2007). The Norwegian positions were communicated at a high political level in Brussels and in central capitals in EU member-states. Thus, the Norwegian government sought to influence the negotiations through informal contact with EU and national representatives (Jevnaker 2014).

Meanwhile, a political process on climate policy had been initiated within Norway. Following a Green Paper from the government (MoE 2007), a cross-partisan compromise on climate policy was adopted by the Storting (the Norwegian parliament) in January 2008 (Stortinget 2008a).[4] According to this 'climate settlement', Norway's 2020 target would be to reduce its emissions by 30 per cent compared with 1990, including international offsetting and domestic forest uptake (MoE 2007, pp. 37, 51). Moreover, agreement was reached on the distribution of cuts at home vs. abroad compared with a business-as-usual scenario for 2020, of which about two-thirds should be carried out in Norway. In practice, however, this entailed that domestic emissions should be reduced by only 12–16 per cent, including forest uptake (roughly half of the targeted 30 per cent), retaining flexibility for the remainder for which cost-efficiency calculations would be decisive.[5] No sector-specific targets were set (Stortinget 2008a, p. 3). Finally, Norway aimed to become 'carbon-neutral' by 2050, entailing a contribution to reducing *global* emissions equivalent to 100 per cent of its *domestic* emissions, compared with 1990 (Stortinget 2008a). This did not say anything about the size of Norwegian emissions in 2050, only that these should be neutralized through Norwegian funding of emissions reductions conducted at home or abroad (MoE 2007, p. 36).

9.4 IMPLEMENTING THE PACKAGE

Norwegian implementation processes have differed from those of EU member-states. Norway would not be bound by the EU's overarching climate and energy-policy goals, but it would have to adopt legislation relevant to the cooperation within the EEA framework. Roughly put, this means legislation on the internal market or on issues already regulated by the EEA Agreement. Moreover, there might be negotiations between

Norway and the Commission on potential 'adjustments' before implementation (MFA 2012, pp. 16–17). Whereas the Commission monitors national implementation of EU legislation for the member-states, this work is conducted for Norway by the ESA. The ESA can initiate infringement proceedings in the case of non-compliance on a particular piece of EU legislation. However, it can only do so once legislation has been added to the EEA Agreement. With the exception of the ESD, the EU expected Norway to implement the climate and energy package.

EU ETS

The Norwegian government was positive to the revision of the ETS Directive, although the Ministry of the Environment was at the time still engaged in implementing the previous version of that Directive. The revised ETS Directive of 2009 was, in fact, implemented early. As part of the implementation process, public consultations were carried out in 2008 (on the proposed Directive) and in relation to the transposition process in 2012 (MoE 2008, 2012a).[6] Stakeholders welcomed the Directive, although several noted that a share of the revenues from auctioning should be earmarked for energy and climate R&D purposes – which did not happen.[7] Moreover, the Environment Agency included affected companies in the process of drawing up a list of companies to receive allowances (NIMs), subject to Commission approval (MoE 2012c).[8]

The transition from the second to the third trading phase saw major changes to Norway's participation in the EU ETS, even beyond the introduction of an EU-wide cap on emissions. First, new sectors were included, like aluminium and ferro-alloys, which are major industries in Norway. Previously, such companies had been regulated mainly through voluntary agreements for GHG reduction (Environment Agency 2014b). As these companies were highly power-intensive and exposed to international competition, the government made use of the possibility under the Directive to adopt measures to shield energy-intensive industries against indirect costs of the ETS: it adopted a compensation mechanism whereby companies would receive money back ex post for ETS-induced increases in power prices (MoE 2013). The mechanism was administered by the Environment Agency, whose decisions on compensation to the individual companies for the first year (2013) triggered protests from industry, dissatisfied with the level of support granted. As a result, the energy-intensives appealed the decisions to the Ministry of the Environment.[9]

Second, harmonized European allocation rules were introduced. The energy-intensives were critical of the Commission's proposed guidelines for allocating allowances to ferro-alloy companies. The allocation was seen as too low. The Commission's proposal was treated in comitology, where representatives from the Commission and member-states would discuss the matter. Norway also participated with observer status here. In the end, the allocation to ferro-alloy companies was adjusted slightly upwards (Climate Change Committee 2011). Nevertheless, as decisions based on the allocation guidelines were made in Norway, energy-intensive companies filed a complaint with the Ministry of the Environment on the Environment Agency's interpretation of the guidelines. The Ministry consulted with the ESA, and, being informed that the guidelines expressed what the Commission would accept in approving a NIM, upheld the Environment Agency's practice.

Third, the petroleum industry was entered on the carbon-leakage list. Companies within this sector had to purchase credits during the second trading phase, as they were not on the carbon-leakage list. For the third phase, this sector was put on the list, and would thus receive free allowances under carbon-leakage provisions. As the petroleum sector contributed to about a quarter of Norwegian emissions, the Norwegian government wanted to avoid weakening the incentives for reducing emissions here. Therefore, the government sought to influence the provisions on carbon leakage (Environment Agency 2011, p. 5). When it became clear that this would not be possible, Norway announced that it would take steps to ensure that the cost of emitting for the petroleum sector would remain constant despite the changes induced by EU legislation (MoE 2011). In 2012 the Ministry of Finance revised the CO_2 tax in order to uphold the cost of emitting for this sector (MoE 2012b, p. 113).

The Norwegian NIM was approved by the Commission in mid-2013 (ESA 2013a). In terms of actual allocation, this had been delayed by a year, as allowances for 2013 were not distributed until the spring of 2014, together with allowances for that year. In 2013, 71 per cent of emissions in traded sectors were covered by allowances, with the remaining 29 per cent covered by purchased credits (Statistics Norway 2014c). Emissions from the traded sectors have grown over time, jumping by 300,000 tonnes CO_2 equivalents in 2013. Two-thirds of this growth has come from the petroleum sector (Environment Agency 2014a).

Effort-Sharing Decision

While Norway was not required to implement the ESD, and was not given a national target for the non-traded sectors, it had to implement transport legislation. Transport is the most important among the non-traded sectors. The EU had adopted transport-specific legislation that related to fuel quality and car emissions. Although this would contribute to emissions reduction within an important non-traded sector, these legal acts were de-linked from the main package (see Chapter 5). Both acts were labelled 'EEA-relevant' from the EU side.

The Fuel Quality Directive was under the competence of the Ministry of the Environment, although the Ministry of Petroleum and Energy was also heavily involved, in view of the impact on the petroleum sector. Public consultations were carried out in 2009 and 2010, but stakeholder input dealt mainly with technicalities. An informant noted that the Ministry of Petroleum and Energy was generally sceptical towards the life-cycle approach incorporated in the Directive. Moreover, the Norwegian government voiced concern over the default values for different fuels, to be adopted through comitology (Norway 2012). Whether a common value or separate default values should be assigned to conventional and unconventional oil was a controversial issue among the EU member-states. Norwegian actors – and Statoil in particular – were concerned about the use of default values in general, irrespective of the issue of unconventional tar-sands (see Chapter 10). A further major concern for Statoil was the reference year against which efforts at reducing emissions from different fuels in a life-cycle perspective would be measured. Statoil stated that it had taken steps to reduce emissions in production *prior* to that year – efforts that would not be taken into account. The Norwegian government sought to get the reference year adjusted (*Teknisk Ukeblad* 2012), but the continuing EU-level discussions on the Directive delayed this process (see Chapter 10). As of September 2015, the Directive had not yet been added to the EEA Agreement, and Norwegian transposition was still pending.[10]

The second transport law was the Car Emissions Regulation (henceforth: Car Regulation). The Norwegian government, with the Ministry of Transport and Communications responsible, was initially positive to the Regulation. Whereas the EU target for 2020 was 95 g/km, the Norwegian target was 85 g/km (Vegvesenet 2014). Moreover, Norway used the tax system to incentivize consumers to purchase less-polluting private cars. The Regulation was not expected to have significant impact on Norway, given its lack of an automobile industry. However, Norway's Public Roads Administration, a government agency subsumed under the Ministry of Transport and Communications, notified the Commission that the technical solutions needed for reporting would not be in place on time.[11] Officially, the government postponed implementation of the Car Regulation until further notice because Norway had no automobile industry. Electric vehicles now made up a relatively large share of new cars (10 per cent), but the share as regards the total car fleet remained below 1 per cent (Statistics Norway 2014a). Average emissions from new passenger cars decreased, on average by 123 g CO_2/km in 2013 – down 7 g from 2012 (Vegvesenet 2014). The Regulation remained unimplemented as of September 2015.

Renewable Energy Directive

Thanks to Norway's decarbonized on-shore power system, the previous RED had come through easily, but the revised Directive extended beyond the power sector, which for Norway entailed higher costs. Moreover, the EU's method for calculating national targets – a combination of GDP/capita and existing renewables share – would mean a much higher target for Norway than the EU average (20 per cent) or the highest member-state target (49 per cent for Sweden). It was expected that the target for Norway would be about 72 per cent, as compared with slightly above 58 per cent in 2005 (Bøeng 2010, p. 50). Increasing the Norwegian renewables share would be costly, and was controversial, so the government wanted a lower target.

The Ministry of Petroleum and Energy, responsible for RED implementation, arranged public consultations in 2008 and 2009 – prior to the EEA process (MoE 2012d).[12] Stakeholders were generally positive to the Directive but were divided as regards the target. Whereas environmental NGOs and windpower interests wanted a high target, power producers saw the need for measures to ensure that additional power production would replace fossil fuels – if not, such production would go to export. Industry was concerned that the Directive could represent an obstacle to industrial growth or to CCS.[13]

In parallel, Norway had been discussing a joint green certificates market with Sweden, a market-based support scheme for incentivizing the expansion of renewables. Although the green certificates scheme had originally been launched as a process separate from RED, they eventually became interconnected (interviews, 2014). As a precondition for entering into a green certificate agreement, Sweden demanded that Norway should implement RED. Otherwise, Sweden would not be able to count Norway's potential contributions to increased renewables production towards its own renewables target, as permitted by the flexible mechanisms allowed for under RED. Sweden was less concerned about the size of a Norwegian renewables target, but was adamant that Norway should implement the Directive. This message did not go unnoticed by the Commission, which understood that Norway would have to transpose the Directive in order to get the certificates scheme.

Thus, the Norwegian government was negotiating with the EU on the Norwegian renewables target, and with Sweden on the overall capacity expansion to be funded through a common green certificates market. In December 2010, agreement was reached (Boasson 2015). Norway had acceded to Swedish demands that about 13 TWh should be funded by each country, resulting in an overall target of 26.4 TWh by 2020.[14] The distribution of actual capacity addition between the two countries would be decided by the market, but each would be credited with half of the overall contribution. Norwegian stakeholders, consulted through a public hearing, welcomed the scheme but were divided on the issue of technology-neutrality vs. differentiated rates for specific technologies (MPE 2011b).

As the green certificates scheme was being finalized, negotiations with the EU on the Norwegian target continued. In June 2011, Norway adopted legislation for technology-neutral green certificates, a scheme that would become operational in January 2012 (Stortinget 2011). In late July, the Ministry of Petroleum and Energy announced that it had reached agreement with the EU on Norway's renewables target: 67.5 per cent, as compared with 58.2 per cent in 2005 (MPE 2011c). The Ministry was not enthusiastic about the target, noting that increasing production could reduce power prices, which would act as a disincentive to energy efficiency. Higher supply without growing demand leads to lower prices, making energy-saving investments less attractive. Norway's plan for achieving its national renewables target was submitted in June 2012, with the green certificates presented as the main instrument. The NREAP did not include any commitments on particular measures within the transport sector (MPE 2012a, p. 13).

The green certificates brought an increase of 6.2 TWh during 2012 and 2013, of which only 0.9 TWh was in Norway (mainly hydropower) and 5.3 TWh in Sweden (mainly windpower, and to a lesser extent biomass) (NVE & Energimyndigheten 2014).[15] As regards progress in renewables expansion within the power sector, then, as laid down in its NREAP, Norway was well above its interim targets for 2011 and 2012, and thus on track to being credited with half the renewables expansion within the joint certificates market. Within the transport sector, however, Norway was increasingly falling behind schedule.[16]

CCS Directive

Norway had high ambitions as to CCS, having launched a national project that the government had compared to a 'moon landing'. Between 2007 and 2012, almost €1 billion was allocated for funding CCS technology development, and together with major Norwegian companies like Statoil, the government established a CCS test centre (MoE 2012b, pp. 112–16). Norway had supported the process towards an EU directive on CCS, emphasizing the major potential for off-shore storage in Norway. However, the resultant CCS Directive included liability provisions for stored CO_2 that had been added towards the end of the EU's legislative process (between October and December 2008), provisions to which the Norwegian government as well as major stakeholders like Statoil were critical (CLIMIT 2013).[17] The public consultation on the draft directive, conducted jointly by the Ministry of Petroleum and Energy and the Ministry of the Environment in 2008, had not included the liability provisions. Nevertheless, the CCS Directive was added to the EEA Agreement in mid-2012, entering into force in mid-2013 (EEA Joint Committee 2012).[18]

While the Ministry of the Environment was responsible for incorporating the environmental aspects of transport and storage of CO_2, responsibility for regulating transport and storage of CO_2 on the continental shelf was delegated to the Ministry for Petroleum and Energy together with the Ministry of Labour – the latter's mandate focused on the safety aspects (health, safety and environment regulation) of transport and storage of CO_2 in such areas (MPE 2011a, p. 25; Stortinget 2013). When the decision incorporating the CCS Directive in the EEA Agreement entered into force in mid-2013, national legislation for implementation in Norway remained absent. The ESA initiated an infringement process by sending a letter of formal notification in September 2013 (ESA 2013b).

Meanwhile, the Norwegian CCS development project was struggling. The CCS test centre experienced problems and repeated delays; and in October 2013 an audit was released, criticizing the government's management of the project (Office of the Auditor General 2013). This occurred shortly after national elections, and the new minority right-wing coalition government announced a re-think on Norway's CCS policy (MPE 2013b).

During the spring of 2014, parallel public consultations were carried out by the Ministry of the Environment and the Ministry of Petroleum and Energy on executive measures for implementing the CCS Directive. While the Ministry of the Environment presented a draft executive measure on storage provisions in general (MoE 2014a), the Ministry of Petroleum and Energy offered a draft measure on transport and use of subsea reservoirs for CO_2 storage (MPE 2014b). The message from the ministries was that the two proposals together would constitute an integrated approach to regulating CO_2 storage. However, several respondents pointed out significant discrepancies between the draft regulations, and stakeholders in general were critical of the Directive's liability provisions, which had been incorporated into the draft measures.[19]

That autumn, the government announced its new policy approach to CCS. Flexibility was emphasized, opening up the possibility of funding projects abroad. Moreover, it was noted that there was limited demand for storage – no capture projects had been set up in Europe (see Chapter 10). As a result, the potential for off-shore storage in Norway in the foreseeable future was downplayed (MPE 2014a). Norwegian business interests supported the new approach (see Federation of Norwegian Industries 2014). Later that year, the ESA reacted to Norway's continued lack of implementation of the CCS Directive by issuing a final warning. Shortly thereafter, both ministries implemented their previously proposed executive measures (MoE 2014b; MPE 2014a).

Conclusions

The various parts of the EU's climate and energy package were transposed separately. First, the ETS Directive was transposed not only correctly but also in line with the EU deadline, despite the accepted time-lag of up to one year due to the additional EEA processes. Emissions-trading enjoyed strong support among politicians, bureaucrats and stakeholders alike. Implementation was conducted in close contact with energy-intensive industries in particular. Second, although Norway was not obliged to implement the ESD, specific transport legislation for reducing emissions from the most important non-traded sector was EEA-relevant, but implementation of both the Fuel Quality Directive and the Car Regulation has remained pending. Third, as regards the RED, Norwegian policymakers opposed a high domestic renewables target, while views among stakeholders were mixed. Despite the protracted negotiations on the national target, this Directive was implemented roughly on time.[20] In implementing the RED, Norway introduced technology-neutral green certificates to support behaviour already established: hydropower has long traditions in Norway and dominates the on-shore electricity system. How new renewable energy should reduce emissions was not addressed, and there were only weak or few instruments to incentivize the shift from fossil fuels to renewables. Fourth, although initially positive to the CCS Directive, Norwegian authorities and stakeholders became embroiled in discussions on its content. The Ministry of Petroleum and Energy and the Ministry of the Environment held overlapping responsibilities, and the major stakeholder Statoil was critical of liability for operators in case of leakage of stored CO_2. The Directive was eventually implemented, but with delays and after pressure from the ESA. Thus, we see that the ETS and RED were implemented correctly and on time, whereas the CCS Directive was implemented with delay.

Changes triggered by the package were by and large absorbed into existing practices rather than having a transformative impact. In application, moreover, behavioural change in terms of reducing emissions on the one hand, and increasing the share of renewables on the other hand, had been relatively low thus far, with little change to be expected towards 2020. A higher carbon price within a reformed ETS, and stricter conditions for receiving free allowances and compensation, could serve to incentivize emissions reduction in the longer term. While expansion of renewable power production is likely to be lower than in Sweden, increased capacity in Sweden will also affect Norway due to the integration of the Nordic power market. Together with greater capacity for trading electricity with EU member-states, this could reduce emissions outside of Norway.[21] Moreover, the trade option might reduce incentives for electrification of transport or petroleum: first, because higher export could bring higher power prices, and second, because it could lower the push for such electrification from the power producers.[22] While the Norwegian government has generally favoured electrifying transport (mostly private vehicles), recent years have seen major discussions highlighting the lack of governmental push – but also the technological and economic challenges – for electrification of the petroleum sector (*Teknisk Ukeblad* 2013). Technology development on electrification or CCS could pave the way for lower emissions towards 2050; but reduced activity within the petroleum sector after 2030 could also reduce emissions. However, both factors are surrounded by significant uncertainty.

9.5 EXPLAINING IMPLEMENTATION

We have noted the mixed pattern of support and resistance to different parts of the EU package. This section analyses the causes of implementation performance for the separate parts of the package, as well as the package as a whole.

Fit and Adaptation Pressure

As outlined in Chapter 2, misfit between the EU and Norway would make implementation of the EU climate and energy package difficult. Such ‘misfit’ could include differences between existing Norwegian policies and the EU package as well as differences in energy use and emissions, influencing the objectives that policies are intended to attain.

As regards the extent of (mis)fit at the level of policies and measures, the ETS Directive was highly aligned with the emphasis within Norwegian climate policy on cost-efficiency and the use of general economic measures applied cross-sectorally. Moreover, the possibility of shielding energy-intensive industries at risk of carbon leakage was also well aligned with established practice in Norway. While the EU's decision to shield the petroleum sector from carbon-pricing challenged Norwegian practice, this was absorbed by making revisions to existing tax regulations (the CO_2 tax), thus maintaining the status quo. Generally, then, the high fit goes far in explaining the high implementation performance for the ETS Directive. With little impact on the status quo, application characterized by low behavioural change was unsurprising.

What, then, of transport legislation as relevant to the ESD? Established Norwegian practice differed from the EU's life-cycle approach under the Fuel Quality Directive, with low implementation performance as expected from a misfit perspective. As to the Car Regulation, it fit well with existing Norwegian practices – but implementation was postponed, and is thus not explained by the misfit perspective.

Moving on to the RED, we note that it had a poor fit with existing climate-policy measures in Norway, and collided with prevalent conceptions of climate policy: it was not seen as cost-efficient or as acting to reduce emissions. The association between the RED and ETS added further turbulence, as the former was seen as undermining the latter. Nevertheless, Norwegian transposition was timely and correct – which is not explained by misfit. In application, then, adaptation pressure was heavily reduced by means of flexible mechanisms (joint implementation with Sweden through green certificates), thereby reducing the push for behavioural change in Norway.

As to the CCS Directive, it generally seemed to fit well with existing CCS priorities in Norway. However, implementation was turbulent and delayed – which is not accounted for by the misfit perspective.

Moreover, the package itself was tailored to a different situation. The baseline situation saw differences between the EU and Norway. First, there were striking differences between the EU average and Norway in terms of energy mix, with the former far more reliant on fossil fuels (78.6 per cent) than the latter (55 per cent) (Commission 2010a). Second, the EU had become increasingly reliant on importing this energy, in contrast to energy-exporting Norway. As a result, the EU was more concerned about security of supply than Norway. Third, the composition of GHG emissions differed, with diverging trends (excl. forest uptake): EU emissions were sinking (–4.8 per cent in 2007 compared with 1990), with the energy sector as the largest emitter (38.2 per cent), predominantly from power and heating (87 per cent). In contrast, Norwegian emissions had risen sharply (+29.3 per cent), although its power and heating sectors were largely decarbonized (Commission 2010b). At the EU level, then, increasing the share of renewables addressed the concern both for climate change and for security of supply – as well as offering a venue for new growth and employment (see Chapter 5). The EU's energy and emissions profile allowed for synergies between climate and energy policy. In Norway, however, no such synergies could be established, as concern for security of supply was low, whereas the share of renewables was already quite high. Rather than focusing on synergies, the prevalent view in Norway was that the renewables target was undermining emissions reduction in general and the ETS in particular.

Summing up, we see that misfit can explain only parts of the package. Implementation performance of the ETS and Fuel Quality Directives was as expected. However, and unexpectedly, misfit did not delay the RED; and troubles arose for the CCS Directive and the Car Regulation despite the good fit. At a more fundamental level, major differences in energy and emissions profiles between the EU and Norway show the difficulties of reproducing synergies from the EU package in Norway. The EU package was the product of a process that effectively reduced the scope for Norwegian influence – and the package had not been adjusted to the situation in Norway.

Domestic Politics

The domestic politics perspective (see Chapter 2) relaxes the assumption that maintaining the status quo is the favoured option, and considers the response from different actors within the state, government and society when accounting for implementation performance.

Governmental actors

From 2005 to 2013, Norway was governed by a government with a parliamentary majority, thereby reducing the role of the Storting. For the package, there were no indications that parliamentary involvement was important. Transposition could also be affected by a change of government, through executive or legislative measures. However, there was no change in government in Norway throughout most of the period studied (2005 to 2013), and we find few indications of any impact on transposition processes as a result of the change to a minority right-wing government in the autumn of 2013.[23] Summing up, parliamentary involvement or change of government emerges as less important in accounting for implementation performance.

Societal actors

The ETS Directive generally saw low costs imposed on target groups in Norway due to carbon-leakage provisions for the petroleum sector and energy-intensives, in combination with a decarbonized power sector. Moreover, higher carbon prices in the EU would favour gas at the expense of coal, which would benefit Norwegian gas-export interests. As regards application, there were no major costs for target groups; and, as expected, no opposition emerged against this Directive as a whole.

Transport, as part of non-traded sectors under the ESD, experienced cost-concentration for the petroleum sector under the Fuel Quality Directive, whereas Norway lacked an automobile industry to be targeted by the Car Regulation. While the continued non-implementation of the former could be understood in light of the costs, the latter case was not explained by this.

Concerning the RED, the implementation of a support scheme – the green certificates market – entailed concentrated benefits to power producers while distributing the costs to consumers. However, power producers had no guarantee that policies would be implemented to allow for increased demand – through export, conversion of fossil-fuel use within the transport or petroleum sectors, or through new industrial production – without which power prices would fall, thereby reducing revenues and asset values. Meanwhile, being major power consumers, the energy-intensive industries were wary of any increase in costs. Although exempt from financing the green certificates and receiving a rebate on grid tariffs, they were concerned about higher costs due to the grid reinforcements triggered by expanding power production.[24] Thus, power producers were ambivalent – welcoming a support scheme but anxious about a price collapse – while energy-intensive industries were concerned about higher power prices. Such a mixed response from societal groups would have been expected to create hurdles for implementation. Nevertheless, implementation was basically on time. In light of the misfit identified in the previous section, this remains puzzling.

The CCS Directive entailed concentrated costs for Norway's petroleum sector, which would be liable for stored CO_2. While this goes far in explaining the delays, it cannot explain why the CCS Directive was eventually implemented.

As to the package as a whole, the costs and benefits for traditional energy-intensive industries were mixed, as their high consumption of renewables-based electricity entailed a lower potential for emissions reduction through their energy consumption. On the other hand, this also meant a lower impact from decarbonization on power prices, as the power sector would not undergo major changes. Moreover, the benefits for Norway's power producers were greater than those of their European counterparts, because they would reap benefits from previous decarbonization while also gaining new prospects for growth through the new support scheme. The costs and benefits from the package for the petroleum sector were peculiar. On the one hand, petroleum companies could have been relieved of the cost of polluting through carbon-leakage provisions (had not the government made changes to other legislation in order to avoid this), and their largely fossil-fuel-based energy consumption would not be taken significantly into account when calculating the Norwegian renewables share.[25] On the other hand, the liability provisions – as seen from the companies' perspective – serve to reduce the incentives for engaging in CCS. While the Fuel Quality Directive might have contributed to some re-balancing, its non-implementation entailed that the overall impact of the package was to reduce incentives for emissions reductions within this sector, which was responsible for more than a quarter of Norway's emissions.

Policy style

As Norway is not an EU member, stakeholder input mainly dealt with issues that would be decided at the domestic level within the overarching framework set by the EU package. While talks between Norway and the EU have tended to be closed due to the need for secrecy in the negotiation situation, the development of domestic policies and measures has been more aligned with 'normal' policymaking, where Norway has a firm tradition of neo-corporatist policymaking.

Beyond public hearings, the ETS Directive saw high involvement of – and attention to the interests of – the energy-intensive industries. This facilitated agreement on compensatory measures for shielding the latter from costs caused by the ETS. Companies within the traded sectors were involved in the government's work on the NIM. With changes being introduced on allocation in particular from the third trading period, the open policy style seems to have been important in avoiding hurdles in implementing this Directive.

With transport legislation related to the ESD, Norwegian authorities were highly attentive to the concerns of the petroleum industry for the Fuel Quality Directive. Public and private actors jointly opposed this piece of legislation, which goes far in explaining Norway's non-implementation. Regarding the Car Regulation, no domestic automobile industry existed that could be included in the process.

For the RED, the neo-corporatist policy style might have delayed the implementation process, given the existence of widespread scepticism, if not outright opposition. Here, the closed negotiations seem to have been crucial in keeping the process on track, especially as the public consultations had shown divisions among stakeholders on important details. The process was governmentally driven, but target-group concerns were attended to at a later stage, including decisions on power export, tax harmonization and exemptions from funding the green certificates (energy-intensives, later also refineries).

While the CCS Directive started out as a more open process, the EU's inclusion of liability provisions shortly before the adoption of the Directive made Norwegian implementation challenging. When the two ministries in charge finally announced draft legislation, this was not welcomed by the petroleum sector – the main target group. Stakeholder responses were not only negative, but also concerned about differences at the level of detail between the two draft regulations. The petroleum industry's lobby group (the Norwegian Oil Industry Association) was critical of the short deadline for providing a response to the hearing (MoE 2014c), an indication of a preference for stronger inclusion in the policymaking process than what had been the case. Moreover, implementation became further delayed. Nevertheless, given the government's limited scope for manoeuvre on the controversial liability issues – and having invested politically in CCS – greater involvement of stakeholders could have resulted in a deadlock.

Administrative organization

Beyond initial assessments in interministerial committees, the different parts were distributed to different ministries in line with their responsibility for the subject matter. They were then implemented separately and according to 'normal' procedure, not through a package approach.

The ETS Directive was handled by the Ministry of the Environment, although the associated revision of the CO_2 tax was the responsibility of the Ministry of Finance. Both were strongly in favour of the ETS, however, and despite the need for coordination in order to tackle changes introduced with the third trading phase (expanded sector coverage, harmonized allocation rules, carbon-leakage status to petroleum sector, revision of the CO_2 tax), the implementation process was efficient.

Among the non-traded 'ESD' sectors, transport legislation was handled by the Ministry of the Environment (Fuel Quality Directive) and the Ministry of Transport and Communication (Car Regulation). The impact of the former Directive on the petroleum sector meant that the Ministry of Petroleum and Energy was also heavily involved. Here, the general scepticism rather than administrative organization seems to have been decisive for the low implementation performance.

The RED was the responsibility of the Ministry of Petroleum and Energy, but here too, the Ministry of Finance played a role. Both ministries were sceptical of the Norwegian renewables target, and the Ministry of Finance became engaged in discussions on the government's negotiation position towards the EU. In the end, however, transposition was on time. As application of the green certificates scheme required coordination between the national and the local level of government as part of the licensing process, this reduced the possibility for the national government to undertake swift expansion of domestic renewables-based power production.

For the CCS Directive, fragmented responsibility at the national level offers more in explaining the implementation process. This was an area of overlap between the Ministry of Petroleum and Energy (responsibility for resource issues) and the Ministry of the Environment (responsibility for environmental issues). Informants identified the Ministry of Petroleum and Energy as the lead ministry, while noting that there had been a 'turf war' between this ministry and that of the Environment, especially regarding off-shore storage of CO_2. While other governmental bodies were affected as well, the competition between these two ministries as to responsibility for the CCS Directive – especially given their diverging views as to how CO_2 storage should be regulated – was a source of delay, despite the Norwegian government's previous vocal support for this Directive. This goes far in explaining the late transposition and low application performance for this Directive.

Summing up domestic politics, we find that the Norwegian Storting and changes of government were less important in accounting for implementation. A neo-corporate policy style facilitated implementation through coordination with societal groups where these shared the government's support of directives (ETS Directive), whereas a more closed process for stakeholder involvement became crucial for avoiding delays (RED) or deadlock (CCS Directive) when target groups were divided or opposed. Diverging views among ministries with shared responsibility also caused delays (CCS Directive), and joint resistance from government and societal groups can explain non-implementation of the Fuel Quality Directive.

Other factors

Some unresolved puzzles remain: neither misfit nor domestic politics can offer a satisfactory account for the high implementation performance regarding the RED. While the closed policy style could explain why the mixed response from target groups did not reduce implementation performance, it is inadequate for explaining the outcome, since important governmental actors were sceptical. Because of the EEA Agreement, however, the cost of non-compliance is quite high. Rejection of EEA-relevant EU legislation may result in cancellation of entire chapters of the EEA Agreement, putting cooperation within a whole policy area on ice. For Norway, this could have severe economic implications, since the EU is its most important market, accounting for 80 per cent of exports. So far, such rejection has yet to occur and, from a Norwegian perspective, the costs of such a step would be very high. Norway's association with the EU through the EEA Agreement can thus explain why the RED was implemented despite the misfit, which the government sought to reduce through flexible mechanisms. This should have been the case also for the Fuel Quality Directive, but delays at the EU level in deciding implementing legislation (default values for calculation of life-cycle emissions) to a greater extent meant that negotiations between Norway and the EU became long-drawn, with high stakes for important Norwegian petroleum interests making it worthwhile. Likewise, continued EU-level negotiations may explain how Norway could postpone EEA incorporation of the Car Regulation, despite the low stakes.

9.6 CONSEQUENCES FOR NORWAY'S 2030 POSITION

Experience with and learning from implementation of the 2020 package was expected to affect Norway's 2030 position (see Chapter 2). The Norwegian experience as regards implementing the 2020 package was mixed – processes ranged from smooth and timely to turbulent and delayed. Norway had a relatively positive experience with the ETS, where it could shield and compensate sectors at risk of carbon leakage. Although the RED was implemented, the Norwegian government was not happy with this Directive and the accompanying national renewables target; and the strict liability provisions for CCS meant a burden. Finally, transport legislation remained unimplemented, with Norway highly critical of various elements of the EU's fuel-quality legislation. As noted, Norway's implementation of the 2020 climate and energy package did not give rise to transformative changes. However, it contributed to raising interest in electricity export, although with a certain ambivalence due to the conflict of interest between power producers and energy-intensive industries.

As regards our analytical framework, a mixed experience was expected to lead to a preference for re-packaging climate and energy policy for

2030 based on those components of the 2020 package with which a country had had positive experiences, and decoupling the difficult parts (see Chapter 2). What position, then, did Norway have on EU climate and energy policy towards 2030? Norway maintained its preference for a single target – with the ETS as the main instrument (Norway 2013b, 2014). It had previously supported ETS reform in order to incentivize investments in low-carbon technology, although it was against limiting the use of international credits (CDM/JI) for complying with the ETS (Norway 2013a). Moreover, Norway opposed specific targets for renewables and energy efficiency: these should be incentivized by the ETS and the internal energy market. Further, policy tools for energy efficiency should be left to the nation-states to decide (Norway 2013b). When it became clear that the EU would retain multiple targets, Norway stressed the importance of coordination among these in order to avoid distorting the internal energy market (Norway 2014).

Moreover, Norway continued to favour CCS, also to be supported by an ambitious ETS. Norway called for making CCS mandatory for new coal-fired power plants, but stressed that transport and storage regulation should be flexible (Norway 2013b, 2014). Finally, Norway re-stated its ambition of at least one capture facility by 2020, although this would not necessarily be located within its borders: instead, Norway would 'play a part' in the construction of a pilot (Norway 2013b). The Norwegian government later indicated an open attitude towards using EEA and Norway Grants to fund CCS in Europe (Norway 2014).

Finally, Norway regarded the EU internal energy market as the major factor for enhancing security of supply. While noting the expanded capacity for electricity exchange between Norway and the EU in 2020,[26] emphasis was placed on the role of gas: Norway underlined the importance of market-based instruments to ensure future investments in gas. With a higher ETS price, moreover, gas could replace coal and support intermittent renewable energy, thereby contributing to emissions reduction (Norway 2013b, 2014). Norway was highly critical of the energy union-related idea of joint purchase of gas, which it saw as distorting the market (Norway 2014).

Thus, Norway wanted re-packaged policies for 2030: a single target, with market-based instruments for reducing emissions (ETS) and ensuring security of supply (internal energy market). For the 2020 package, Norway had, as noted, had positive experiences with the ETS, in contrast to the turbulence surrounding the national renewables target and the liability provisions for CCS. Re-packaging based on positively experienced elements was in line with the expectation stated in Chapter 2, although Norway continued to support CCS despite the controversy surrounding the liability provisions and the difficulties in setting up a pilot project in Norway. However, the latter was reflected in the government's position, which stressed the need for flexible regulation of transport and storage, and ambitions for a domestic pilot had been lowered.

Subsequent developments did not alter the main picture, with one notable exception. In early 2015, Norway announced its climate ambitions for 2030: a 40 per cent reduction in emissions, meant to correspond to the EU's 40 per cent target adopted in October 2014 (MoE 2015b; see also Chapter 10). However, the latter's emphasis on 'domestic' emissions reduction was expected to entail a collapse of the system for international credits (CDM), as most demand had come from the EU. Access to such credits had been a crucial element of Norwegian climate policy, without which the country would face far higher costs in reducing its emissions. Norway's 40 per cent target was made conditional on access to flexible mechanisms, not only within the traded sectors but also within non-traded sectors; and the government announced that it would seek to enter into an agreement with the EU – separate from the EEA Agreement – on reducing emissions from the non-traded sectors (MoE 2015b).

Summing up, Norway's 2030 position underlines and expands its interconnectedness with the EU. An important motivation for Norwegian participation in the ESD was access to flexible mechanisms for the non-traded sectors. Although the impact of a domestic ESD target in Norway still remains to be seen,[27] the 2020 package did not change Norway's basic position, which remained stable: emphasis on market-based instruments with flexibility to enable domestic emissions to be offset.

9.7 CONCLUSIONS

Although Norway is not an EU member, it has committed to implementing EU legislation of relevance to the EEA Agreement. Beyond the important energy-trade relations, Norway cooperates extensively with the EU on climate and energy issues. Nevertheless, Norway is inadequately represented within EU policymaking. The EU's package approach to climate and energy policy for 2020, with heavy involvement of the European Council, reduced Norway's possibilities for influence. This proved particularly challenging because of the significant differences between Norway and the EU as regards energy (energy mix and energy-trade balance) and emissions (sector profile and overall trend). Sustainability and energy-security objectives could not be combined as at the EU level – thereby reducing the scope for reproducing synergies and side-payments foreseen at the EU level in Norway.

This chapter has shown a mixed pattern regarding Norway's implementation of the 2020 package: some parts of the package went more smoothly (ETS Directive) than others (RED, CCS), and there has been continued non-implementation of package-associated transport legislation (Fuel Quality Directive and Car Regulation). This mixed pattern can be explained through a combination of factors: differences between the situation of Norway and that of the EU, domestic politics and characteristics of the EEA Agreement – with the added note that foot-dragging in implementation is easier if discussions are still ongoing at the EU level.

The changes triggered by Norway's implementation of the 2020 package were not transformative – they were absorbed. Moreover, access to flexible mechanisms within the ETS Directive and RED, as well as carbon-leakage provisions, reduced the pressure for behavioural change in Norway. Thus, the 2020 package was no game-changer for Norway as regards decarbonization, although the consequences of the Fuel Quality Directive – if implemented – could have been more far-ranging.

Generally, Norway's experiences with the 2020 package had little impact on its 2030 position. The emphasis on a single target and the ETS remained stable from the 2020 package to the later 2030 process. However, changes in EU priorities – notably towards 'domestic' emissions reduction – forced Norway to look in new places for possibilities to offset domestic emissions, reconsidering its stance on participation in EU climate policy for non-traded sectors. Beyond that, however, the EU's 2030 framework was fairly well aligned with Norwegian preferences, with relatively more weight given to the climate target at the expense of the targets for renewable energy and energy efficiency (see also Chapter 10). That being said, however, there are no guarantees that prospective package legislation will not include elements that may cause headaches in Norway.

NOTES

1. The chapter is based on document analysis as well as interviews, conducted between May and September 2014, in Norwegian. A list of interviewees can be found at the end of the book. See also Chapter 2.
2. In 2012, the amount of natural gas exported was equivalent to ten times the total Norwegian 'normal year' production of electricity (120 TWh), of which less than 20 TWh at most has been exported per year (MPE & NPD 2013, p. 44; Statistics Norway 2014b).
3. This challenge goes beyond climate and energy policy, as highlighted by an official report commissioned by the government (NOU 2012, pp. 170, 860).

4. This was supported by all political parties except the right-wing populist Progress Party.
5. It was estimated that Norway could deduct 3 mtoe CO_2 equivalents from domestic emissions in the 2020 target under the Kyoto Protocol due to absorption of CO_2 from forests (Environment Agency 2014b, p. 7). See Jevnaker (2014) for a more detailed account of the figures on the Norwegian 2020 targets.
6. Stakeholder responses can be found at MoE (2008).
7. As budget issues are outside the EEA Agreement, Norway was exempted from Article 10 (3) of the Directive, which includes a recommendation to use income for low-carbon investments.
8. Formally, the ESA makes the decision.
9. In 2015, some complaints were upheld, whereas others were rejected (MoE 2015a).
10. A draft EEA decision was reported to have been sent to the Commission in July 2013 (Europalov undated).
11. The requirements would necessitate changes in the Public Roads Administration's existing system. Although they would not require additional costs, as a revision of the system was already underway, more time would be needed (Norway 2010).
12. Stakeholder submissions can be found in MoE (2012d).
13. Increased power consumption through new industry installations would reduce the renewables share, while CCS projects would require an increased use of fossil fuels (e.g. gas-fired power plants), which as such – irrespective of carbon capture – would reduce the overall renewables share.
14. Sweden had a target of adding 25 TWh by 2020, of which roughly 12 TWh was expected to be realized by the end of 2011.
15. Although market based, competition was slightly skewed: In Sweden, different technologies received differing rates of support, while the Norwegian system was technology-neutral. The latter had been necessary in order to get the Norwegian Ministry of Finance to accept the scheme. Moreover, the tax regime was more favourable for windpower in Sweden, due to high depreciation rates.
16. The overall renewables share had grown to 66.1 per cent in 2012 (up from 64.8 per cent in 2011), well above the interim targets (62.6 per cent in 2011, 62.8 per cent in 2012). Within the transport sector, the figure had remained fairly stable (from 4.1 per cent in 2011 to 4.6 per cent in 2012), and was below the interim targets (4.6 per cent in 2011, 5.3 per cent in 2012) (MPE 2012b, p. 15; MPE 2013a, p. 2).
17. The disagreement also touched on the sensitive issue of whether the geographical scope of the EEA Agreement extends to the continental shelf, on which the EU and Norway disagree, which had been put firmly on the agenda when the EU adopted the Offshore Safety Directive in 2013.
18. This delay was caused by Icelandic policy-processes.
19. For stakeholder input, see MoE (2014c).
20. Implementation is deemed 'timely' if an EU legal Act is incorporated into the EEA Agreement *and* necessary national legislation has been adopted on time (max. one year later than EU deadline).
21. If Norwegian/Nordic renewable-based electricity replaces electricity produced from fossil fuels in, say, Germany or the Netherlands, this will count towards reducing emissions in the latter countries.
22. Without export, increasing supply by adding new production capacity could act to reduce prices. Power producers would be interested in avoiding lower prices – which mean lower profits for them – by pushing for policies that can stimulate new demand. Electrification of transport and petroleum could increase demand for electricity.
23. At the level of detail, the new government proposed a price floor for the compensation mechanism for energy-intensives under the ETS. Due to heavy criticism across the board – most notably from business and trade unions but also from the opposition – this was withdrawn.
24. In late 2014, the government proposed exempting refining installations (petroleum sector).

25. Roughly put, the share is calculated by dividing overall renewables production and renewables consumption by overall energy consumption. Energy consumption within energy-producing sectors is more or less excluded from the latter figure. Thus excluded from the calculation is the use of natural gas within the petroleum sector, which would have lowered the Norwegian renewables share significantly (Bøeng 2010). This was already part of the Directive, and was not a particular concession to Norway.
26. New interconnectors are to be built to Germany and the UK.
27. For non-traded sectors, the EU target was 30 per cent reduction as compared with 2005, with domestic targets ranging between 0 and 40 per cent (according to burden-sharing among member-states) to be deduced from this. The Norwegian Minister for Climate and the Environment expected the Norwegian target to be located at the higher end of this spectrum. Negotiations with the EU were expected to follow after the EU's own process on non-traded sectors, to be formally launched in 2016 (MoE 2015b).

REFERENCES

Boasson, E.L. (2015), *National Climate Policy: A Multi-field Approach*, London: Routledge.

Bøeng, A.C. (2010), 'Konsekvenser for Norge av EUs fornybardirektiv', *Økonomiske analyser*, 4, 48–58.

Climate Change Committee (2011), *Climate Change Committee – 46th Session: Summary Record* (CLIM B1/Lp Ares(2011) 466608), Brussels: European Commission.

CLIMIT (2013), 'Kommersielle aktører nøler med CO2-lagring', 15 January, *CLIMIT*, accessed 3 December 2014 at http://www.climit.no/no/Sider/Kommersielle-akt%C3%B8rer-n%C3%B8ler-med-CO2-lagring.aspx.

Commission (2010a), *Europe's Energy Position – Markets and Supply: Market Observatory for Energy*, Brussels: European Commission.

Commission (2010b), *EU Energy in Figures 2010: CO2 Emissions by Sector, Extended Time Series*, accessed 11 November 2014 at http://ec.europa.eu/energy/publications/doc/statistics/ext_co2_emissions_by_sector.pdf.

EEA Joint Committee (2012), *EØS-komiteens beslutning nr. 115/2012 av 15. juni 2012 om endring av EØS-avtalens vedlegg XX (Miljø)*. (2012/EØS/56/12). Brussels: EEA.

Environment Agency (2011), *Regler om tildeling av vederlagsfrie kvoter for perioden 2013 til 2020*. Report TA-2826, Oslo: Klif.

Environment Agency (2014a), *Økte klimagassutslipp fra kvotebedrifter*, Oslo: Environment Agency, accessed 5 December 2014 at http://www.miljodirektoratet.no/no/Nyheter/Nyheter/2014/Mai-2014/Okte-klimagassutslipp-fra-kvotebedrifter.

Environment Agency (2014b), *Kunnskapsgrunnlag for lavutslippsutvikling*, Report M229, Oslo: Miljødirektoratet.

ESA (2013a), *Allocations of Greenhouse Gas Emission Allowances in the EEA EFTA States Cleared*, PR(13)66, Brussels: ESA.

ESA (2013b), *Letter of Formal Notice to Norway for Failure to Implement Directive 2009/31 on Geological Storage of Carbon Dioxide*, Brussels: ESA. On file with author.

Europalov (undated), *Drivstoffdirektivet (endring): kvaliteten til bensin og dieselolje – reduksjon av drivhusgasser*, accessed 8 December 2014 at http://europalov.no/rettsakt/drivstoffdirektivet-endring-kvaliteten-til-bensin-og-dieselolje-reduksjon-av-drivhusgasser/id-374.
Europaportalen (2007), *Veikart for fornybar energi*, accessed 3 December 2014 at http://www.regjeringen.no/nb/sub/europaportalen/eos/eos-notatbasen/notatene/2007/mars/veikart-for-fornybar-energi.html?id=523737.
Eurostat (2007), 'Energy: yearly statistics 2005', *Eurostat Statistical Books* (2007 edition), Brussels: Eurostat.
Federation of Norwegian Industries (2014), 'Brev til SMK – regjeringens arbeid knyttet til CO2-fangst, transport og lagring (CCS) – anbefalinger', accessed 15 October 2015 at http://www.norskindustri.no/siteassets/dokumenter/annet/2015-02-12-regj-arbeid-knyttet-til-ccs.pdf.
Jevnaker, T. (2014), *Norway's Implementation of the EU Climate and Energy Package: Europeanization or Cherry-Picking?*, FNI Report 7-2014, Lysaker: Fridtjof Nansen Institute.
MFA (2012), *EØS-avtalen og Norges øvrige avtaler med EU: sentrale prioriteringer og virkemidler i norsk europapolitikk* (Meld. St. 5 (2012–2013)), Oslo: Ministry of Foreign Affairs.
MoE (2007), *Norsk klimapolitikk* (Meld. St. nr. 34 (2006–2007)), Oslo: Ministry of the Environment.
MoE (2008), *Høring av EU-Kommisjonens forslag til direktiv som endrer kvotedirektivet (2003/87/EF) for å forbedre og utvide EUs kvotesystem*, Oslo: Ministry of the Environment, accessed 15 October 2015 at https://www.regjeringen.no/nb/dokumenter/horing-av-eu-kommisjonens-forslag-til-di-2/id544029/?regj_oss=10.
MoE (2011), *Felles EU-regler for tildeling av kvoter*, Press release, Oslo: Ministry of the Environment, accessed 15 October 2015 at http://www.regjeringen.no/nb/dokumentarkiv/stoltenberg-ii/md/Nyheter-og-pressemeldinger/pressemeldinger/2011/felles-eu-regler-for-tildeling-av-kvoter.html?id=667815.
MoE (2012a), *Endringer i klimakvoteloven* (Prop. 68 L (2011–2012)), Oslo: Ministry of the Environment.
MoE (2012b), *Norsk klimapolitikk* (Meld. St. 21 (2011–2012)), Oslo: Ministry of the Environment.
MoE (2012c), *Klimakvoteloven endret,* Oslo: Ministry of the Environment, accessed 31 October 2013 at http://www.regjeringen.no/nb/dokumentarkiv/stoltenberg-ii/md/Nyheter-og-pressemeldinger/pressemeldinger/2012/klimakvoteloven-endret.html?id=683690.
MoE (2012d), *Høring: direktiv for å fremme bruk av fornybar energi*, Oslo: Ministry of the Environment, accessed 13 June 2015 at https://www.regjeringen.no/nb/dokumenter/horing–direktiv-for-a-fremme-bruk-av-f/id544629.
MoE (2013), 'Forskrift om CO2-kompensasjon er vedtatt', accessed 31 October 2013 at http://www.regjeringen.no/nb/dokumentarkiv/stoltenberg-ii/md/Nyheter-og-pressemeldinger/nyheter/2013/forskrift-om-co2-kompensasjon-er-vedtatt.html?id=736302.

MoE (2014a), *Høring: utkast til nytt kapittel i forurensningsforskriften om miljøsikker lagring av CO2 samt mindre endringer i avfallsforskriften og konsekvensutredningsforskriften* (høringsbrev), Oslo: Ministry of the Environment, accessed 15 October 2015 at http://www.regjeringen.no/nb/dep/kld/dok/hoeringer/hoeringsdok/2014/Horing-Utkast-til-nytt-kapittel-i-forurensningsforskriften-om-miljosikker-lagring-av-CO2-samt-mindre-endringer-i-avfallsforskriften-og-konsekvensutredningsforskriften/Horingsbrev.html?id=754221.

MoE (2014b), *Forskrift om endring i forskrift om begrensning av forurensning (forurensningsforskriften) og forskrift om gjenvinning og behandling av avfall (avfallsforskriften)*, Oslo: Ministry of the Environment.

MoE (2014c), *Høring: utkast til nytt kapittel i forurensningsforskriften om miljøsikker lagring av CO2 samt mindre endringer i avfallsforskriften og konsekvensutredningsforskriften*, accessed 5 August 2015 at https://www.regjeringen.no/no/dokumenter/Horing-Utkast-til-nytt-kapittel-i-forurensningsforskriften-om-miljosikker-lagring-av-CO2-samt-mindre-endringer-i-avfallsforskriften-og-konsekvensutredningsforskriften/id754217.

MoE (2015a), *Klageavgjørelser om CO2-kompensasjon*, Oslo: Ministry of the Environment, accessed 10 April 2015 at https://www.regjeringen.no/nb/aktuelt/klageavgjorelser-om-co2-kompensasjon/id2405243.

MoE (2015b), *Ny utslippsforpliktelse for 2030 – en felles løsning med EU, Ministry of the Environment*, Oslo: Ministry of the Environment, accessed 15 June 2015 at https://www.regjeringen.no/nb/aktuelt/ny-utslippsforpliktelse-for-2030/id2397859.

MPE (2006a), 'Støtteordning for fornybar elektrisitet' (Press release 117/06), Oslo: Ministry of Petroleum and Energy, accessed 15 October 2015 at http://www.regjeringen.no/nb/dokumentarkiv/stoltenberg-ii/oed/Nyheter-og-pressemeldinger/pressemeldinger/2006/stotteordning-for-fornybar-elektrisitet.html?regj_oss=1&id=271970.

MPE (2006b), 'Støtteordningen for elektrisitetsproduksjon fra fornybare energikilder' (Press release 135/06), accessed 15 October 2015 at https://www.regjeringen.no/no/aktuelt/stotteordningen-for-elektrisitetsproduks/id437126.

MPE (2007), 'Nye samtaler med Sverige om grønne sertifikater til fornybar energi' (Press release 173-07), Oslo: Ministry of Petroleum and Energy, accessed 15 October 2015 at http://www.regjeringen.no/nb/dokumentarkiv/stoltenberg-ii/oed/Nyheter-og-pressemeldinger/pressemeldinger/2007/nye-samtaler-med-sverige-om-gronne-serti.html?regj_oss=1&id=493551.

MPE (2011a), *Fullskala CO2-håndtering* (Meld. St. 9 (2010–2011)), Oslo: Ministry of Petroleum and Energy.

MPE (2011b), *Høring – lov om elsertifikater: høringsuttalelser*, Oslo: Ministry of Petroleum and Energy, accessed 5 December 2014 at http://www.regjeringen.no/nb/dep/oed/dok/hoeringer/hoeringsdok/2010/horing–lov-om-elsertifikater/horingsuttalelser.html?id=630808.

MPE (2011c), *Mål om norsk fornybarandel på 67,5 prosent i 2020* (Press release 67/11), Oslo: Ministry of Petroleum and Energy, accessed 15 October 2015 at http://www.regjeringen.no/nb/dokumentarkiv/stoltenberg-ii/oed/Nyheter-og-pressemeldinger/pressemeldinger/2011/mal-om-norsk-fornybarandel-pa-675-prosen.html?id=651715.

MPE (2012a), *Fakta 2013: energi- og vannressurser i Norge*, Oslo: Ministry of Petroleum and Energy, accessed 15 October 2015 at https://www.regjeringen.no/no/dokumenter/fakta-2013/id712168.

MPE (2012b), *National Renewable Energy Action Plan under Directive 2009/28/EC: Norway*, Oslo: Ministry of Petroleum and Energy, accessed 15 October 2015 at https://ec.europa.eu/energy/sites/ener/files/documents/dir_2009_0028_action_plan_norway__nreap.pdf.

MPE (2013a), *Member State Progress Reports under Directive 2009/28/EC*, Oslo: Ministry of Petroleum and Energy.

MPE (2013b), 'Ønsker ny vurdering av CO2-håndtering', accessed 2 November 2014 at http://www.regjeringen.no/nb/dep/oed/pressesenter/pressemeldinger/2013/onsker-ny-vurdering-av-co2-handtering.html?id=744870.

MPE (2014a), *Proposisjon til Stortinget (forslag til stortingsvedtak) for budsjettåret 2015* (Prop. 1 S (2014–2015)), Oslo: Ministry of Petroleum and Energy.

MPE (2014b), *Utkast til Forskrift om transport av CO2 og utnyttelse av undersjøiske reservoarer på kontinentalsokkelen til lagring av CO2*, Oslo: Ministry of Petroleum and Energy.

MPE & NPD (2013), *FACTS 2013: The Norwegian Petroleum Sector*, Oslo/Stavanger: Ministry of Petroleum and Energy, and Norwegian Petroleum Directorate, accessed 15 October 2015 at http://npd.no/en/Publications/Facts/Facts-2013.

Norway (2010), Cached version of 'CO2 krav til personbiler' [content removed in a 2014 revision], *Europaportalen*, accessed 7 December 2014 at http://web.archive.org/web/20120924045627/http:/www.regjeringen.no/nb/sub/europaportalen/eos/eos-notatbasen/notatene/2007/des/co2-krav-personbiler.html?id=553584.

Norway (2012), 'Norwegian comments concerning the Commission proposal for specification of the methodologies for calculation of greenhouse gas emissions for fuel and energy under article 7a in the Fuel Quality Directive (Directive 98/70/EC, as most recently amended by directive2009/30/EC)', accessed 15 October 2015 at http://www.regjeringen.no/upload/EuropaPortalenImages/2905/Norwegian_comments__2_.pdf.

Norway (2013a), 'EUs karbonmarkedsmelding', *Europaportalen*, accessed 7 December 2014 at http://www.regjeringen.no/nb/sub/europaportalen/eos/eos-notatbasen/notatene/2013/feb/eus-karbonmarkedsmelding.html?id=716347.

Norway (2013b), *Letter to the Commissioners for Climate Action, and Energy* (19 December), Oslo: Ministry of Foreign Affairs, Ministry of the Environment, Ministry of Petroleum and Energy. On file with author.

Norway (2014), 'Norwegian position on the proposed EU framework for climate and energy policies towards 2030' (6 October), accessed 15 October 2015 at http://www.regjeringen.no/upload/UD/Vedlegg/Protokoll/141006-posisjonspapir-EU-klima-energi.pdf.

Norwegian Oil Industry Association (2010), *Norway's Petroleum History*, accessed 13 August 2013 at http://www.norskoljeoggass.no/en/Facts/Petroleum-history.

NOU (2012), *Utenfor og innenfor: Norges avtaler med EU* (2012:2), Oslo: Official Norwegian Report.

NVE & Energimyndigheten (2014), *Et norsk-svensk elsertifikatmarked: Årsrapport 2013*, Oslo/Stockholm: Norwegian Water Resources and Energy Directorate, and Swedish Energy Agency.

Office of the Auditor General (2013), *Riksrevisjonens undersøkelse av statens arbeid med CO2-håndtering*, Oslo: Office of the Auditor General.

Statistics Norway (2013), *Statistikkbanken*, accessed 19 December 2013 at https://www.ssb.no/statistikkbanken.

Statistics Norway (2014a), *Registrerte kjøretøy, 2013*, accessed 5 December 2014 at http://www.ssb.no/bilreg.

Statistics Norway (2014b), *Electricity, Annual Figures, 2012*, accessed 8 December 2014 at http://www.ssb.no/en/energi-og-industri/statistikker/elektrisitetaar/aar.

Statistics Norway (2014c), *Utvidet kvoteplikt fra 2013*, accessed 27 October 2014 at http://www.ssb.no/natur-og-miljo/artikler-og-publikasjoner/utvidet-kvoteplikt-fra-2013.

Stortinget (2008a), *Avtale om klimameldingen,* Oslo: Stortinget, accessed 15 October 2015 at https://www.regjeringen.no/contentassets/fbe5a5829a5d468fab6e4eec0a39512d/avtale_klimameldingen_2008_01_17.pdf.

Stortinget (2008b), *Møte i Europautvalget mandag den 1. desember 2008 kl. 16.15*, accessed 3 December 2014 at https://www.stortinget.no/nn/Saker-og-publikasjoner/Publikasjoner/Referat/Europautvalget/2008-2009/081201.

Stortinget (2011), *Vedtak til lov om elsertifikater* (Lovvedtak 61), Oslo: Stortinget.

Stortinget (2013), *EU/EØS-nytt – 3. oktober 2013*, accessed 4 November 2013 at http://www.stortinget.no/no/Hva-skjer-pa-Stortinget/EU-EOS-informasjon/EU-EOS-nytt/2013/EUEOS-nytt - 3-oktober-2013.

Teknisk Ukeblad (2012), 'Norske raffinerier kan straffes for klimatiltak', accessed 29 July 2013 at http://www.tu.no/olje-gass/2012/02/24/norske-raffinerier-kan-straffes-for-klimatiltak.

Teknisk Ukeblad (2013), 'Edvard Grieg-plattform ikke forberedt på elektrifisering tross krav fra myndighetene', accessed 7 December 2014 at http://www.tu.no/petroleum/2013/12/17/edvard-grieg-plattform-ikke-forberedt-pa-elektrifisering-tross-krav-fra-myndighetene.

Tjernshaugen, A. and O. Langhelle (2011), 'CCS som politisk lim', in J. Hanson, S. Kasa and O. Wicken (eds), *Energirikdommens paradokser: innovasjon som klimapolitikk og næringsutvikling*, Oslo: Universitetsforlaget, pp. 172–94.

Vegvesenet (2014), *Gjennomsnittlige CO2-utslipp*, accessed 3 December 2014 at http://www.vegvesen.no/Kjoretoy/Fakta+og+statistikk/Kjoretoy+og+drivstoff/Klimagassutslipp/Gjennomsnittlige+CO2-utslipp.

Wicken, O. (2011), 'Kraft fra infrastruktur til marked', in J. Hanson, S. Kasa and O. Wicken (eds), *Energirikdommens paradokser: innovasjon som klimapolitikk og næringsutvikling*, Oslo: Universitetsforlaget, pp. 126–44.

10. Comparative analysis and consequences for EU 2030

Jon Birger Skjærseth, Per Ove Eikeland, Lars H. Gulbrandsen and Torbjørg Jevnaker

What comparative lessons can be drawn from domestic implementation of the EU's climate and energy package for 2020? And what are the implications for the negotiations on 2030 targets and policies? Chapters 6 to 9 have analysed the implementation of the EU climate and energy package in Germany, the Netherlands, Norway and Poland. This chapter begins by comparing whether and how these countries have implemented the package with a view to the EU 2020 targets and beyond. The second part seeks to explain variation in implementation. In the third part, we explore whether experiences with implementation have affected national positions on new EU 2030 policies, and how positions fed into the reform of the EU climate and energy framework for 2030 agreed by the European Council in October 2014. Part four discusses the consequences of the new 2030 framework and related policies for a low-carbon transformation.

10.1 COMPARING IMPLEMENTATION

Three parts of the package required legal transposition: the revised ETS Directive, the CCS Directive and the RED. And here, three observations stand out. First, all four countries eventually transposed these Directives in national legislation, although with lengthy delays in some instances. Second, the Netherlands has generally been most timely, whereas Poland has had the most severe delays.[1] Polish transposition of the ETS took place alongside systematic legal opposition to the Commission's interpretation of the Directive as a measure for contributing to low-carbon solutions. The RED transposition deadline was 5 December 2010 but Poland did not notify of its transposition until 2015, and uncertainty about new legislation remains. Poland's transposition of the RED was characterized by draft proposals, consultations, amendments and new draft proposals. Among the most contentious issues were whether national subsidies should apply also to less mature technologies than co-firing biomass and coal, and for small power companies, as noted in Chapter 7. Third, transposition of the CCS Directive has been severely delayed in all four countries examined here except the Netherlands.

Application concerns the adoption of domestic policies and measures and associated behaviour as regards the 2020 targets. Here we can note good overall progress for the EU as a whole as regards achievement of the 20–20–20 by 2020 targets (European Environment Agency 2014). By 2013, total GHG emissions were already 19 per cent below 1990 levels, seven years ahead of the target date. The share of renewables in total energy consumption had reached 15 per cent by 2013 (Eurostat 2015a). Energy consumption had decreased, although still not sufficiently to achieve the 2020 energy-efficiency target (Commission 2014a). Progress was uneven, with significant variation across member-states, as shown by our four case studies.

To a large extent, application of the ETS Directive has involved strategic responses made by industrial firms, as no specific national targets have been set for emission reductions in the ETS sectors. Industrial installations receive individual amounts of allowances per year that add up to an annual cap set at the EU level. The Directive instructs the ETS sectors to cut their emissions by 21 per cent by 2020 compared with 2005; as of early 2014, emissions were already down 19 per cent (European Environment Agency 2014). The reductions from 2008 reflected mainly down-scaled activity levels because of the economic crisis, import of external credits and fuel-switching to renewable energy because of RED implementation (European Environment Agency 2014, p. 28).

Member-state governments must calculate the number of free allowances to be allocated to each installation in their territory, in accordance with the rules set out in the Benchmarking Decision (NIMs) and notify the Commission of these by 30 September 2011. For Germany, NIMs were found to be not entirely compliant with the allocation rules and were rejected, causing delays in allocation. All the NIMs combined showed that national preliminary calculations exceeded the maximum amount of allowances available in 2013, whereupon the Commission decided to undertake correction by a similar factor across all sectors.[2] This intervention shows that the Commission served as a 'watchdog' for ensuring implementation in line with goals. Another national government policy task, limited to the CEECs, was reporting on investments to restructure the power sector – a condition for this sector to receive temporary free allowances. The Commission temporarily rejected Poland's reported investment plan, which had included new coal-based generation capacity.

The ESD did not require legal transposition, but national policies and measures targeting the non-ETS sectors to reach the national target. Again, we note a good track record for the EU as a whole as regards meeting the 10 per cent reduction target by 2020 compared with 2005. In 2014, 25 member-states were following trajectories necessary for meeting their 2020 targets (European Environment Agency 2014). Projections indicate a large surplus of AEAs by 2020 (European Environment Agency 2014). The economic crisis has contributed significantly to this situation by reducing activity in the non-ETS sectors, transport in particular (Eurostat 2015b).

Progress has differed for the countries studied here. Poland belongs to the group of ten countries that were allowed to increase their emissions because of their low GDP/capita. Poland was the only one of these where emissions by 2013 were clearly above the trajectory needed to meet the 2020 target. The Netherlands and Germany are among the 18 member-states that had to reduce their emissions. For 2013, emissions in the Netherlands were well below the trajectory towards 2020, whereas Germany was one of only two member-states with emissions above the trajectory (European Environment Agency 2014, p. 51).[3] Transport-sector emissions represent the largest part of non-ETS emissions. They were significantly higher in Germany in the 2010–2013 period compared with the previous five-year period, partly as a result of inadequate policies and measures (European Environment Agency 2014).

In the end, all countries are likely to attain their national reduction targets by 2020 because of surplus in the AEA 'market' and various flexibility mechanisms provided for in the ESD. First, the Netherlands has transferred additional activities or substances from the non-ETS sector to the ETS sectors, thereby reducing its national ESD target. The Netherlands has also used the CDM option. Second, all member-states can bank, borrow and transfer AEAs, within certain limits.

Application of the RED has also been mixed for our four case countries.[4] All have implemented a wide range of policies and measures to promote renewables, with subsidy schemes funded by consumers as the main policy measure. Norway and Poland have relied on technology-neutral green certificate systems that stimulate the cheapest technologies – mainly hydropower (Norway), and co-firing coal and biomass (Poland). The Netherlands has a feed-in premium scheme. Since 1990, Germany has had a feed-in tariff system that provides generous support and guaranteed market uptake for a range of renewable electricity sources.

The system was continued under Germany's implementation of the RED until 2012, and then gradually replaced by a mixed market premium/ feed-in tariff system.

The share of German renewables in total electricity consumption increased from 14 per cent to 27 per cent between 2007 and 2014. However, the RED target is measured as share of renewables in gross final energy consumption. Here, the figures for Germany are less impressive, because of slow uptake of renewables in the heating and transport sectors. From around 7.5 per cent in 2007, the share of total renewables in German energy consumption increased, from 10.4 per cent in 2010 to 12.4 per cent in 2013 – still some distance from the RED target of 18 per cent by 2020. This is comparable to Poland, whose share of renewables increased from 9.2 per cent in 2010 to 11.3 per cent in 2013. Poland may very well reach its lower 15 per cent target for 2020 due to its significant co-firing of biomass and coal. Norway is on track to attaining its 67.5 per cent RED target as a result of the green certificate market with Sweden: the share increased from 61.2 per cent in 2010 to 65.9 per cent in 2013, although new capacity is predominantly built in Sweden. The renewable energy situation in the Netherlands is less favourable. The share of renewables was only 3.7 per cent in 2010 and 4.5 per cent in 2013, still far from its EU target of 14 per cent by 2020. The Netherlands has adopted new policies and measures but appears unlikely to achieve its 2020 target. Concerning the RED's specific 10 per cent renewable transport fuel target, most of the EU member-states have come about halfway to their targets. Progress has been sluggish in all our four country cases (Eurostat undated).

The CCS Directive covers legal rules for safe carbon capture and storage and does not oblige the member-states to implement and realize CCS. It should, however, in combination with subsidies for construction provided by revenues from sale of ETS allowances through NER 300, provide greater legal and financial certainty for investors to realize the up to 12 CCS pilot plants by 2015 stipulated by the European Council in 2007. In this case, we refer to 'application' as realization according to targets agreed by the European Council.

Germany, the Netherlands, Norway and Poland all had plans for developing new CCS plants partially funded by EU money. None of them (or indeed any of the EU-28) has done so, or probably will by 2020.[5] Poland for example, has cancelled all CCS pilot projects. In Europe, only two large-scale plants are operating, both started up in Norway before the Directive was adopted, and intended for enhanced oil recovery off-shore (Commission 2014b; see also Chapter 4).[6]

Table 10.1 summarizes implementation of the core package in the selected countries.

Table 10.1 Implementation of the package by 2014 with a view to 2020

	ETS		ESD		RED[3]		CCS	
	Transposition	Application[1]	Transposition	Application[2]	Transposition	Application	Transposition	Realization
Germany	Yes	On track	–	Not on track	Yes	On track	Yes, but delayed	No
Netherlands	Yes	On track	–	On track	Yes	Not on track	Yes	No
Norway	Yes	On track	n.a.	n.a.	Yes	On track	Yes, but delayed	No
Poland	Yes	On track	–	Not on track	No	Partly on track	Yes, but delayed	No

Notes:
[1] On track as regards implementation of NIMs and the EU 21 per cent GHG reduction target in the ETS sectors between 2005 and 2020. To a large extent, the EU ETS is a harmonized EU-level instrument.
[2] Situation as of 2013.
[3] Not including the specific 10 per cent renewable transport target.

Implementation with View to the Long Term

The EU package of climate and energy policies has not as yet proven itself a collective 'game-changer' by shaping policy conditions for long-term low-carbon transformation. First, our four case-study countries have planned very differently for the 2050 horizon. Germany has since 2010 set ambitious decarbonization targets aimed at ultimately establishing a low-carbon economy by 2050 (at least 80 per cent reduction in its GHG emissions). The Netherlands and Norway have adopted carbon-neutrality strategies that differ from Germany's in committing only parts of the effort to be domestic, pledging instead investments abroad. Emphasis is placed on international flexible mechanisms rather than domestic transformation. Poland has not adopted any long-term decarbonization or carbon-neutrality goals for 2050.

Second, although assessments indicate substantial EU-level progress towards the 2020 targets, our implementation studies show that the massive cut in GHG emissions thus far in the implementation period masks deficiencies in the policies and behavioural change necessary for long-term transformation. This observation corresponds with the findings of previous research (Dupont and Oberthür 2015). Around half the reductions reflect activity-dampening effects of the economic crisis more than forceful policies towards a low-carbon future (Commission 2014c). Concerning the ETS sectors, which account for about 45 per cent of EU GHG emissions, the drop in activities and emissions reduced demand for allowances and led to a high surplus of 2.1 billion allowances in 2013, expected to increase to 2.6 million by 2020 (Commission 2014d). The carbon price has since around 2011 been too low (less than 10 euros per tonne) to provide real incentives for companies to invest in low-carbon solutions for the future. By contrast, higher carbon prices in the past (around 30 euros in 2008) had in fact led companies to steer their investments in a low-carbon direction, on the expectation that prices would remain high (Skjærseth and Eikeland 2013). Will such previous strategic changes be discontinued? That depends partly on countermeasures linked to the reform of the EU ETS for the fourth trading phase, 2021–2030 (see section 10.4).

In the non-ETS sectors, implementation challenges can be illustrated by the situation of the transport sector, which has had the highest share of total EU GHG emissions – 19 per cent in 2012 (Eurostat 2015c). The EU climate and energy package addresses emissions reductions in road transport through the national ESD targets, the RED, the FQD and the regulation of emissions from new cars, as explained in Chapter 5. After peaking in 2007, transport emissions fell by 10 per cent in the following five years – mainly because of lower volumes of passenger and freight transport, reflecting the economic downturn and the hike in fuel prices (Eurostat 2014). Road transport is responsible for the bulk of transport emissions (71.9 per cent in 2012); also here, emissions have gone down because of the economic crisis and higher fuel prices – a trend that may reverse, with economic growth picking up again and the fall in international oil prices that came in 2014. Promoting energy-efficient and renewable-energy-based road transport remains a crucial unresolved challenge for transforming the transport sector in line with the EU low-carbon transport roadmap target of 60 per cent reduction by 2050 compared with 1990 levels (Commission 2015b).

Also the RED has had limited transformative effect in our selected countries. Poland, Norway and the Netherlands have implemented subsidy schemes to achieve the 2020 targets, but currently have no plans for extending these subsidies beyond 2020. In Poland and the Netherlands, the use of renewables in final energy consumption is based largely on co-firing biomass and coal. Co-firing does not contribute to 'lock out' coal from the energy mix; and the renewables component may prove short-lived if technology-neutral subsidies end before 2020 and coal becomes cheaper than biomass. Poland's long-term strategy involves meeting energy demand through indigenous fossil fuels for as long as possible. Norway is a special case, since hydroelectric power already accounted for nearly 100 per cent of inland electricity production before the EU package. Thus far, the Norwegian/Swedish green certificate market has served mainly to stimulate investments in Sweden. Still, the RED contributes to increasing overall power production, which together with new interconnectors allows for higher electricity export.[7]

Germany is the only example of a 'radical' transformation towards renewable *electricity* production. However, this started not as a result of EU policies, but because of national support policies that were consolidated by the EU's RED and advanced after the Fukushima nuclear disaster.[8] Policy changes in 2014 cut the level of support and brought investments to a lower tempo that may still challenge Germany's ambitious long-term goal of 80 per cent renewable electricity by 2050. Among the challenges are how to balance a system with such a high level of intermittent production without too much coal power; how to avoid overly high costs that may erode popular support; and how to achieve transformation also in the heating and transport sectors. However, massive cuts in technology and installation costs, as well as stable high popular and political commitments to date, indicate that the transformation will be continued.

The post-2020 challenges for the EU as a whole are even more massive. Much hinges on the political willingness of the 28 member-states to continue adopting 2030 policies that benefit renewables as opposed to the continued use of fossil fuels – a challenge made even more daunting with the recent fall in fossil-fuel prices (see section 10.4 and Chapter 11).

CCS was launched by the EU as a long-term low-carbon solution attractive also for the fossil-fuel industry, coal in particular. However, this solution has almost disappeared now, with pilot projects cancelled throughout Europe. According to the Commission, there is a genuine and urgent need for CCS to enable decarbonized power production and energy-intensive industries by 2050 (Commission 2014b). Nearly full decarbonization of industrial processes away from coal will entail enormous costs and technological challenges, but may still be technologically possible through energy efficiency, the use of renewable power, and renewable methane or hydrogen (German Energy Agency 2014).

Conclusions

We must conclude that the EU climate and energy package has not yet managed to set the four countries studied here on the route towards low-carbon economies in line with the EU's long-term decarbonization goal. All four have eventually transposed the required policies, with Poland lagging behind – but application of policies and measures remains challenging. The variance in implementation for 2020 is striking as regards different EU policies and among countries. All are on track with the ETS – which is, however, not functioning as intended. The picture is mixed concerning the ESD and RED – the four countries are fully, partly or not on track towards 2020 targets. We have not found any necessary relationship between transposition and application of the package towards 2020 targets. For example, the Netherlands fully transposed the RED on time but has been lagging far behind its 14 per cent target, due to insufficient policies and measures. Conversely, Poland has struggled to transpose the RED, but may very well attain its 15 per cent target as a result of policies to stimulate co-firing coal and biomass. All four eventually transposed the CCS Directive but no new CCS pilot plants have been constructed.

In a long-term view, the only clear example of radical transformation is the increase of renewable energy in German electricity production. National goals and strategies towards 2050 have been stimulated by EU policies and targets, but with wide variance. Germany has adopted decarbonization as a goal, and aims for GHG emissions reductions of at least 80 per cent. The Netherlands and Norway have adopted carbon-neutrality strategies to be achieved only partly by transformation at home, relying on international flexibility for emissions reduction abroad. And Poland has not adopted any 2050 decarbonization or carbon-neutrality goals.

10.2 EXPLAINING VARIATION IN IMPLEMENTATION

How can variation in implementation be explained? Our first explanation examines variation in 'fit' and adaptation pressure between EU requirements and national status quo. We then turn to variation in domestic politics.

Fit and Adaptation Pressure

The first explanation holds that varying fit and adaptation pressure between EU policy requirements and national status quo can account for variation in transposition and application. High match with pre-existing policies, energy-economic situation, negotiation position and the final EU outcome is expected to increase the feasibility of policy implementation.[9] However, we find that the match between expectations and observations based on this proposition has been low, according to our assessment of implementation of the four core pieces of the package in our case countries (a total of 15 observations, as the ESD is not applicable to Norway) (see Table 10.2).

First, we note that there is a good fit between EU policy requirements and the national situation in 11 of the cases. Two observations can be categorized as 'medium' and two as 'poor'. In Germany, the fit is good for all four policies – corresponding with pre-existing policies, the energy-economic situation and concessions given to Germany during the 2008 negotiations. The Netherlands also has good overall fit. Poland displays more variation: good fit with CCS, poor with the ETS, and medium as to the ESD and RED. The ETS 'punishes' coal the hardest (coal has a 90 per cent share in Polish electricity consumption) and CCS provides an opportunity to combine coal with GHG reductions. Moreover, the negotiated EU package did not meet Polish demands. Norway did not participate in the EU negotiations, but the fit with CCS and the EU ETS was very good. On the other hand, the fit with the RED was poor, as there was no strongly expressed goal of increasing the already high share of renewables in final energy consumption in Norway.

Second, as to implementation, only seven cases can be characterized as 'on track/realized' (see Table 10.1). Four of these are for implementation of the revised EU ETS Directive, which did not necessitate any significant domestic efforts for Phase III, as the National Allocation Plans required in Phase I and II were scrapped.

Third, if we relate fit to implementation, we find a strong match between expectations and observations for only five cases (Table 10.2). All four countries are on track concerning implementation of the ETS, but fit varies. However, Poland is a borderline case with firm but unsuccessful resistance to transposition, as we expect for a country characterized by low fit. Only one case – Germany – is fully in line with expectations concerning the RED. To a significant extent this Directive was a result of German 'uploading' of existing domestic policies to the EU level, as noted in Chapter 6. Concerning the ESD, however, our observations match expectations poorly. The different national ESD targets based on GDP/capita fit well with national situations and were accepted without much resistance. However, the Netherlands is the only country thus far where implementation is on schedule, partly as a result of the flexibility mechanisms in the ESD.

The CCS Directive is of particular interest, as all four countries were enthusiastic about the EU's goals with this technology. CCS fit well with domestic policies and storage capacities, Poland's high reliance on coal, Norwegian and Dutch petroleum production, and German coal. However, transposition of the Directive has been severely delayed in Germany, Poland and Norway. Moreover, realization of CCS has failed on all counts.

Table 10.2 Fit and adaptation pressure: correspondence between expectations and observations

	ETS	ESD	RED	CCS
Germany	Intermediate	Intermediate	Strong	Weak
Netherlands	Strong	Strong	Intermediate	Weak
Norway	Strong	n.a.	Weak	Weak
Poland	Strong	Intermediate	Intermediate	Weak

The EU package approach and conclusions

Part of our investigation throughout Chapters 6 to 9 involved exploring the impact of a policy-package approach. With several EU policies negotiated simultaneously as a package in 2008, issue-linkages at EU level might lead to varying fit and adaptation pressure among the legislative package components. When unanimity is required, member-states can be expected to trade concessions among policies, according to their differing national priorities. Varying fit and adaptation pressure would thus lead each member-state to respond to and implement the components differently.

Issue-linking and 'horse-trading' during negotiations take place behind closed doors and may be challenging to investigate. Lack of transparency may contribute to explain why we can offer relatively few observations of this phenomenon. Or it could be that the package was designed and attuned to varying national situations based on effort-sharing and common impact assessment (see Chapter 5). Bearing this caveat in mind, we can conclude that whether several EU policies are negotiated simultaneously or in isolation may make a difference to 'fit' and thereby help to explain variation in implementation between policies. This is illustrated most clearly in the case of Poland, which agreed to accept a more ambitious RED target than preferred, partly in exchange for concessions on the revised ETS. We have also seen that implementation of the RED in Poland has proven extremely challenging.

Thus, we find that the 'fit' and adaptation pressure approach can only partly explain variation in implementation of the 'core' package in Germany, the Netherlands, Norway and Poland. We have also seen that high fit does not correspond to delayed transposition of the CCS Directive. That leaves several questions: Why has Poland been reluctant on renewables and failed to realize CCS as intended by EU goals and legislation, when energy security and coal are among its major political concerns? Why did Norway struggle to implement the apparently suitable CCS Directive – but manage to stay on track towards its renewables target, despite a high misfit? Why was the Netherlands pleased with the EU's package approach, while implementation of key components has proven difficult? And why has implementation of CCS and ESD in Germany proven challenging, given that the country pushed for the package at the outset? To these and other questions we now turn.

Domestic Politics

The second explanation is that variation in implementation may be due to variation in domestic politics. Differing responses to agreed-upon EU policies may be explained by the state, or government itself, society and the relationship between the state and society where domestic institutions channel influence through various channels. This domestic politics perspective focuses more on internal political and administrative processes.

As noted in Chapter 2, incoming governments may introduce new priorities after EU policies have been negotiated and adopted. Societal actors may be strengthened or weakened by new policies, and may gain or lose political influence. Policy style may vary according to different ways of linking state and societal actors, and administrative organization may be more or less equipped for swift EU policy implementation.

First, the *governments*. Here we simply expect that continuity in (majority) government over time – from the EU 2008 negotiations to national implementation – will improve the likelihood of implementation, given initial consent to the policy in question. The reasoning here is that implementation is affected by the conformity between EU requirements and political preferences. New majority governments that were not involved in the preceding EU policy negotiations may come to veto these policies and the measures needed for implementation.

Our four cases support this proposition only partly. Poland and Norway have been characterized by political stability from 2007 to 2014 and 2013, respectively.[10] The same governments have negotiated the package (in the case of non-EU Norway, followed the negotiations) and have had responsibility for implementation. This indicates that implementation challenges have not been caused by shifts in political priorities as a result of change in government.[11] Poland is a particularly interesting case concerning the revised ETS. The same Polish government agreed to the revised ETS Directive – and then later systematically, albeit unsuccessfully, opposed transposition of the trading system. The Tusk-led Civic Platform did not change its position on the issue. A reasonable interpretation is that the Polish government had accepted an unwanted outcome in 2008.

By contrast, the Dutch case lends support to the expectation that change in government affects implementation. In 2010, the Dutch centre-right coalition minority government replaced the fourth Balkenende cabinet that had been in power from 2007. The new government reduced climate and energy ambitions by lowering the targets for the level of GHG emissions and renewable energy. The new government did not oppose the EU climate and energy package, but RED implementation problems can partly be traced back to lower governmental priority, with less ambitious targets, fewer implementation efforts and changes in regulatory frameworks. The binding EU targets and policies served as a 'floor' for the new government's desire to reduce ambitions.

The German case also gives some support to the proposition. In 2009, the grand coalition government of the CDU/CSU and SPD that had negotiated the package was replaced by a new coalition government of CDU/CSU with the FDP as minority party, in power until 2013.

Implementation of measures to strengthen the ETS was postponed under this government because of internal disagreements. After the SPD again became a minority party in the government after 2013, the German government took a supportive position on the revised EU ETS. However, under Germany's federal political system, *Länder*-level governments also have formal co-legislative powers in federal policymaking through the Federal Council. Voting by the Federal Council had a very strong influence on EU policy implementation in Germany – which proved decisive for transposition delays with the CCS Directive and in curtailing policies and measures relevant for the ESD.

Governments do not transpose or apply EU Directives in a vacuum. They rely on, and are influenced by, *affected societal actors* that may support or resist their goals and priorities. We expect that concentration of costs and distribution of benefits will increase resistance to governmental policies and reduce the likelihood of implementation. Resistance will be particularly difficult to overcome if negatively affected societal actors represent 'social blocs'.

Our cases partly support this expectation. Costs mainly concern fossil-fuel interests (as intended), while the benefits are widely distributed to society at large. The failure of CCS has reinforced the long-term decarbonization costs for fossil-fuel interests. We find that the anticipated costs matter considerably as regards the way affected societal actors respond to policies – the costs they expect if climate policies are tightened in the future to reach the EU 2050 decarbonization target. Direct carbon costs have proved significantly lower than expected, because of the economic crisis. Accordingly, the indirect costs from higher power prices due to the ETS for energy-intensive industries have been low; moreover, they can be compensated according to EU state-aid rules. This option has been used by six EU member-states, including the Netherlands and Germany.[12] Norway has also made use of this option. Costs from renewables in the form of various types of support schemes are generally distributed to society at large through mark-up on customer energy bills. Energy-intensive industries as large consumers have increasingly been exempted from paying the mark-up – notably in Germany.

In Poland, the major affected societal actors are the coal-based energy groups, which also form a 'social bloc' with significant political influence. This industry will be hit hardest by a high carbon price and an increase in decentralized renewables. It sees the EU's push for low-carbon energy technologies as a threat to its future existence, in the absence of CCS. Benefits from new 'green' jobs or energy-technological innovation are not deemed viable opportunities, except for biomass.

Resistance from these major energy groups is an important explanation of ETS opposition and the slow implementation of the RED in Poland.

The highest actual costs of German policies have been experienced by the country's major utilities, whose costs have risen because of the rapid growth in renewable electricity, reducing revenues from and asset values of their conventional plants. The utilities prefer the ETS as a technology-neutral measure that would allow them to switch fuels according to variation in market prices and profitability. They have lined up in support of the EU ETS as the only policy instrument for driving the transition at the German and EU levels. They have long opposed binding goals for renewables and the national feed-in tariff scheme, but have gradually had to accept these measures. Despite free allowances and generous rebates on the feed-in surcharge, energy-intensive industries have opposed strengthening the EU ETS by backloading and the German renewable energy transition. In contrast to the case of Poland and also to our expectations, heavy opposition from industry has not prevented implementation of highly ambitious renewable energy policies in Germany. Thus, we find no automatic relationship between opposition from powerful target groups and implementation challenges.

In the Netherlands, the main sectors targeted are the petroleum industry and energy-intensive industries. Costs are concentrated to these sectors, as well as to Dutch fossil-fuel electricity producers (mainly gas-fired power plants), whereas the benefits are widely distributed. The major Dutch industries see few benefits from a low-carbon economy, and this has delayed application of the RED in particular. As noted in Chapter 8, a pertinent example is the lobbying by British–Dutch petroleum company Shell to undermine EU renewable energy targets since October 2011. This position is shared by electricity producers in the Netherlands. The close connection between the German and Dutch electricity markets has resulted in oversupply of cheap electricity to the Dutch market and reduced profits for Dutch electricity producers. Moreover, unlike other comparable countries, the Netherlands lacks a substantial 'green' technology industry, and there are relatively few 'green' jobs.

In Norway, the distribution of costs and benefits has been mixed. The RED costs were borne by households and medium-sized energy consumers. Benefits for the power industry came with a flipside: power prices would decrease unless additional production could be consumed domestically or exported. Still, the power sector did not oppose the RED, and it supported the ETS. Norway's energy-intensive industries are mainly shielded against direct and indirect carbon costs and have cautiously supported the ETS Directive. They were able to avoid additional costs from the RED, but remained quite sceptical, fearing increased export of electricity, which could bring up power prices. The economically important petroleum industry was exposed to a CO_2 tax before the ETS, and carbon costs would not increase as a result of the revised EU ETS. Indeed, Norway's petroleum industry would benefit from a higher carbon price, which would make gas competitive to coal in Europe. Moreover, this sector's energy use would not be taken into account under the RED. Involvement in off-shore CCS for oil recovery would be affected by the CCS Directive, however, and the petroleum sector opposed the liability provisions of the CCS Directive, delaying transposition. On the whole, there were no clear 'losers' or 'winners' involved for most sectors in Norway.

These cases show that opposition or support from affected societal actors has not necessarily translated into political influence. Influence depends on many factors, including how societal groups are included – formally and informally – in decision-making. And here we expect that a consensual *policy style* founded on open access and broad participation will improve the feasibility of policy implementation, possibly at the expense of the level of ambition.

On this point, our observations are inconclusive. The German case is of particular interest, as Germany is known for broad-based inclusion of societal actors in policymaking and voluntary agreements with industry. The utilities were unhappy with the RED and have largely excluded themselves from the decentralized renewable 'revolution' in German electricity production. The country's traditionally strong utilities have lost their once-privileged status in policymaking, with the growing renewable energy industry gaining leverage at the federal and state (*Länder*) levels. This is closely linked to how the federal political system distributes influence in decision-making to the *Länder* level. The renewable energy industries enjoy better access than do the utilities to policymaking at state level because of their role in creating value-added, employment and tax revenues.

Very different is the situation in Poland, where the style of involvement in climate and energy policies can be placed somewhere between a consensual and imposition style. Formal consultations based on broad-based participation take place regularly, but 'privileged insiders' are given special treatment informally – as has been the case with the protracted implementation of the RED. Poland's major coal-based power companies have had direct access to and influence on decision-making, whereas the renewables industry and the green movement have been partly excluded.

The Netherlands is well known for its policy style based on extensive participation and use of various types of voluntary agreements or covenants. This consensual style can also be seen in the implementation of EU climate and energy policies. In 2013, nearly all societal actors and relevant state institutions, altogether more than 40 organizations, adopted the Energy Agreement, aimed at achieving EU and Dutch climate and energy objectives for 2020. The 2013 Agreement enhances societal support for implementing climate and energy policy by strengthening legitimacy. It may promote implementation – but will most likely prove insufficient for achieving the Dutch RED target.

As for Norway, its policy style in this area can be described as being between Poland and the Netherlands. Norway traditionally provides institutionalized rights of participation for NGOs in all phases of governmental policy. As regards the ETS, this policy style was followed, with extensive involvement of mainly supportive societal groups, which facilitated compensation and the expeditious implementation of the Directive. Other parts of the package departed from the consensual tradition, however. The RED implementation process was closed off by the government. Contrary to our expectation, closing the process as regards partly reluctant stakeholders may well have proved vital for Norway's timely implementation of this Directive: the diverging views among stakeholders – had these been included – would probably have slowed down the process. The CCS implementation process was also closed in Norway. The government had actively propagated the CCS Directive in Brussels, but the petroleum industry opposed implementation of the liability provisions that had been included in the Directive in late 2008. Exclusion of the petroleum industry may have helped Norway in implementing measures in 2014. Nevertheless, the petroleum sector cannot be ignored in Norwegian policymaking – and that point can help to explain the eventual outcome: implementation was delayed, taking place only after warnings from the ESA.[13]

Implementation problems may arise even when governments have willingness to follow through, when affected societal actors are included and their interests are aligned with governmental priorities. One important reason is vertical and horizontal fragmentation in *administrative organization*. 'Fragmentation' refers to the distribution of competence between public authorities and government levels involved in implementation. Here our basic assumption is that supportive or opposing bureaucratic interests with relevant competence can affect implementation. We expected that fragmentation within the government apparatus itself or among various levels of government would tend to reduce the feasibility of implementation.

Our four cases are partly in line with this expectation. Poland appears to be the most fragmented case in terms of laws and regulatory agencies in climate and energy policies. Its late transposition of the CCS Directive was mainly a result of horizontal administrative and legal fragmentation. The Polish government and power companies were genuinely willing to transpose the Directive swiftly, so as to provide investors with legal certainty; and public opposition was low. However, the wrangling among various administrative actors served to protract implementation. Vertically, the local level has been particularly relevant as regards implementation of the RED. Administrative barriers in the form of fragmented laws are a key problem, particularly those concerning impact assessment and licence requirements. In addition, the Polish support system for renewables does not match local administrators' enthusiasm for decentralized solar and wind, particularly in the west of the country.

The Netherlands can be placed at the other end of the spectrum from Poland. Only two ministries had lead responsibility in implementing all the package components, which appears to have facilitated the smooth and quick Dutch transposition process. Concerning vertical application, however, local competence and opposition have obstructed plans from the central government. Local opposition blocked a CCS pilot – whereupon the Dutch government decided not to pursue storage of CO_2 on land in any form, but to focus solely on off-shore CCS. Also on-shore windpower development has been delayed by local opposition and lengthy planning processes. The case of the Netherlands has shown how concentration of public authorities has facilitated transposition of the EU climate and energy package, but local competence and opposition have interfered with application of the RED and CCS.

Norway seems situated in-between Poland and the Netherlands. Two ministries – for Energy and for the Environment – were each responsible for implementing two of the three Directives, both of which were transposed on time. However, their shared responsibility for the CCS Directive resulted in additional delays due to turf war and differing views on how storage should be regulated, further fuelled by the Energy Ministry's traditional receptivity to the preferences of the petroleum industry. Local government played a role in application of the RED in that local considerations fed into the licensing process, with appeals concerning windpower projects in particular. On the other hand, with the RED cooperation mechanism, new installations in Sweden will also be counted towards Norwegian application, and that has weakened the link between local preferences and actual implementation.

In Germany, overlapping ministerial responsibilities created tensions and delays in implementation that were resolved in 2013 by gathering responsibilities for renewable energy policy and energy-market policies in one ministry. Vertical fragmentation in the German administration has posed a great challenge for implementation of parts of the package. German federal administrations share with state (*Länder*) and municipal administrations competencies as regards implementation of federal-level policies. State and municipal governments enjoy wide prerogatives in national planning (spatial planning and local area planning) and have considerable influence on whether final permits will be granted when it comes to, for instance, the construction of new energy facilities. Implementation of the CCS Directive is a clear example. Here, local governments delayed not only the transposition of the Directive but also its later realization, halting CCS projects by preparing *Länder*-level laws permanently banning storage.[14]

The EU package approach and conclusions

How has the EU's package approach of linking different policies affected domestic implementation? Here we conclude that the specific configuration of the package – its policy content consisting of simultaneously EU agreed and linked policies – has affected and facilitated implementation only to a limited extent. The main reason seems to be that the various components of the package have largely been implemented in isolation, and for at least two reasons. First, different policies have followed different tracks in the EU, and policy content and timing have varied, as explored in Chapters 4 and 5. Secondly, different domestic administrative actors have been in charge of implementing 'their' policies largely in isolation from 'others'' policies. Administrative compartmentalization and turf war have obstructed coordination between climate policies and energy policies. In Norway and the Netherlands, the climate and energy administrations were compartmentalized before the adoption of the EU package. The Commission was able to overcome this division with the climate and energy package, but linking these policies domestically has proved more challenging.

There are several nuances to this main observation. We can note that the package approach has had some impact on implementation. In Poland, the EU climate and energy package improved *administrative* cooperation between the Ministry of the Environment and Ministry of Economy, which had competence over climate and energy policy, respectively. Ironically, this improved administrative coordination also made Poland more unified in its opposition to the EU climate and energy package. In Germany as well, these two ministries experienced extended coordination as implementation progressed. Unlike in Poland, however, the German ministries appeared to achieve agreement on a new model for supporting the continued long-term national energy transition and the continuation of ambitious EU energy and climate policies.

Concerning *affected societal actors*, we asked whether a package approach could level out costs by distributing the burden between various domestic sectors. The link between EU state-aid rules and domestic compensation for indirect ETS costs has proved important for reducing opposition from energy-intensive industries, particularly in Germany and the Netherlands. Another important link has been between the EU package and the international climate regime, analysed in Chapters 4 and 5. The ETS, ESD and FQD allow for import of external credits in order to lower the costs of compliance.

With regard to *policy style*, we set out to explore whether a climate and energy package approach could create new constellations of public and private actors and affect access through national consultation processes. In Poland, we find that the RED actually led to a new department within the Ministry of Economy in 2012, giving the renewables industry and 'green' groups a new channel into decision-making. In the Netherlands, the EU climate and energy package may have contributed to the use of the 'Polder model' as regards climate and energy policies. The Energy Agreement has resulted in a new society-wide 'alliance' – comprising virtually all interest organizations in the ETS and non-ETS sectors – that favours a more ambitious climate and energy policy. In Norway, the package has departed slightly from traditional ways of consensual consultations, with consultation processes remaining fairly closed as long as negotiations with the EU have been ongoing. The package has also exacerbated the conflicts of interest between the power industry and energy-intensive industry as to export or internal consumption of surplus electricity. In Germany, a system of wide consultations had already been instituted, with quite universal opportunities for actors to influence policies; and the federal political system included checks and balances for any one of these groups to be given fully privileged status in policymaking. This is illustrated by the increase in leverage of renewable energy interests against the historical background of the high leverage enjoyed by utilities and big industry in German policymaking.

Thus, we conclude that domestic politics may complement 'fit' and adaptation pressure, but appear to explain a more significant share of variation in implementation. Table 10.3 specifies domestic politics-mechanisms that have explanatory power in our cases.

Fragmented administrative organization and opposition from differently affected societal actors emerge as particularly valid in explaining implementation challenges across the cases. All four countries have

Table 10.3 Domestic politics explanations of implementation problems [1]

	Germany	Netherlands	Norway	Poland
Governmental change	X	X		
Negatively affected societal actors		X	X	X
Policy style based on narrow access				X
Fragmented administrative organization	X	X	X	X

Note: [1] Chief explanations. Implementation problems include delayed transposition and application/realization not on track.

experienced implementation challenges due to vertical or horizontal fragmentation within the state apparatus. Opposition from negatively affected societal actors has proven important for all countries except Germany, where opposition from powerful societal actors – power companies in particular – did not block implementation. Policy style based on narrow or 'privileged' access to decision-making has proven least valid across the cases. In Norway, for example, a rather closed implementation process actually facilitated timely implementation. We have not found any specific combinations of factors that serve to enable or inhibit implementation in all four countries according to specific policies: factors interact differently in different national contexts. Norway implemented the undesired RED as a result of distribution of costs and benefits that did not produce any clear 'losers' (and subsequent strong opposition), with a concentration of administrative competence and closed decision-making. The main explanation for Germany's difficulties in implementing CCS and the ESD and success in renewable electricity production can be traced back to a combination of distribution of competencies among federal, state and local levels and varying access to decision-making for different interests. Poland's opposition and reluctant transposition of the ETS and RED have been due mainly to opposition from negatively affected societal actors with privileged access to decision-makers who are hesitant as regards climate policies, reflecting low societal demand. Dutch challenges in implementing the RED can be explained by change in government, reluctant societal actors and vertical administrative fragmentation: local opposition and decentralized competence.

10.3 POLICY REFORM: A NEW EU 2030 CLIMATE AND ENERGY FRAMEWORK

Here we explore the reform of the 2020 policy package in two steps. First, we examine whether implementation experiences had consequences for national positions on the new 2030 climate and energy framework, and then turn to how the varying experiences of the four case countries fed into the EU reform of the 2020 package for 2030.

Consequences for National Positions on New EU 2030 Policies

As explained in Chapter 2, we expect positive experiences with implementation – in the form of combining climate and energy concerns, wider synergies with other national policy objectives, and conflict reduction through compensation and side-payments – to lead to support for higher ambitions through essentially the same set of linked policies and issues. Negative experiences will stem from limited combinations of concerns, synergies and compensation and other domestic implementation challenges, and lead to opposition to strengthening the package approach. Mixed experiences are likely to result in preferences for re-packing as regards specific components of the package.

Poland clearly fits into the 'mainly negative experience' category. It was not pleased with the package, which did not fit well with the Polish energy-economic situation, climate policies or public demands for new policies. The synergies envisaged by the EU as regards 'green' jobs, innovation and keeping coal carbon-neutral through CCS have failed to materialize. Quite the contrary: climate and energy concerns have increasingly clashed. Carbon-pricing and renewable subsidies are seen as threatening coal, thereby curtailing national energy security. Implementation of the RED has proven extremely politically contentious. Growth in the share of renewables has done little to reduce Poland's energy-import dependency: instead, import dependency has increased alongside increase in renewables (Eurostat 2007, 2014). Dissatisfaction with the EU package came to a head in 2012, when the main opposition party to the government presented – albeit unsuccessfully – a resolution in the parliament calling for renegotiation of the EU climate and energy package (Skjærseth 2014).

These experiences resulted in systematic opposition to more ambitious long-term EU policies, as we had expected. Poland vetoed the EU's low-carbon and energy roadmaps for 2050 and opposed 'backloading' of surplus ETS allowances. It also opposed new EU 2030 targets and policies on renewables and energy efficiency, arguing that one single climate target could be adopted after the international climate negotiations in Paris 2015. Poland did not support the idea of 'the EU as leader by example' internationally; and CCS was now seen as impeding the development of clean-coal technologies (Ministry of the Environment [Poland] 2013).

Norway and the Netherlands fit in the 'mixed experience' category. They remained positive to the ETS and recognized that the EU would have to deal with the problem of surpluses. They also remained positive to CCS: it was downplayed, but still important to Norway, which even urged the EU to continue to encourage CCS explicitly in the 2030 framework. As to the RED, Norway did not want it, and the Netherlands seems unlikely to achieve its target. Energy-import dependency is not a concern in Norway, a huge net-exporter. The Netherlands and Norway have not experienced any significant new 'green' jobs as a result of the RED. In line with our expectations, both countries favoured re-packing of 2020 policies for 2030 (European Council 2014a; Norway 2014), and argued that the package should be replaced with one ambitious climate target, with the EU ETS as the core policy instrument.

Germany comes close to the third 'mainly positive experiences' category. It has succeeded in protecting its energy-intensive industry from the competitive disadvantages that could result from carbon-pricing and has benefited greatly from its ambitious renewable policy in terms of 'green' employment and energy technological innovation and diffusion. The package was partly a result of German 'uploading' of existing policies and ambitions. However, we should note several nuances to this generally positive pattern. First, Germany faces challenges with renewables in sectors other than electricity production, and there have been significant costs for small and medium-sized consumers – so it will probably be cautious about promoting an overly-ambitious renewable energy target. Second, there have been challenges in the non-ETS sectors, concerning transport in particular. Third, Germany's experience with renewables has had consequences for neighbouring countries. Poland, for example, has had problems with grid instability as a result of the German expansion of renewables (Piria et al. 2014). And other countries may fear contagion from Germany in terms of problems for their utilities. Nevertheless, Germany has favoured strengthening more of the same 20–20–20 policies based on the RED, the ESD and the ETS, while remaining generally silent on CCS – partly as a result of negative implementation experiences due to local opposition (EurActiv 2014b).

Table 10.4 shows that there is a good overall correspondence between implementation experiences and positions on new policies for the longer term. At the more detailed level, however, we can note some patterns that indicate a negative relationship between position on further policies and whether these policies have been effective in changing behaviour on the ground. The ETS has not worked as intended, but all case countries (except Poland) have lined up for this instrument in the future. CCS has failed completely, but Norway and the Netherlands support this instrument and technology for the future. Conversely, the binding national target of the RED has served to increase the share of renewables, but three of the countries have opposed this instrument for the future, whereas Germany has insisted on continuing with a robust RED. A reasonable interpretation is that countries will resist policy measures that are (perceived to be) costly to implement. Or it might be that increase in renewables has not yet proven effective in reducing energy-import dependency.

Table 10.4 Implementation experiences and long-term positions

	Experience from implementation	Expected position	2030 position
Poland	Mainly negative	Opposition	Opposition: preference for a single climate target, to be adopted after Paris
Netherlands	Mixed	Re-packaging	Preference for a single climate target, to be based on the EU ETS
Norway	Mixed	Re-packaging	Preference for a single climate target, to be based on the EU ETS
Germany	Mainly positive	Strengthening more of the same policies	Internal division, but strengthening more of the same policies

Table 10.5 illustrates two main points: the changes in the share of renewables in the total energy mix of the four countries and the EU; and

the development in energy-import dependencies. Except for the Netherlands, all case countries have seen growth in the share of renewables. However, this has done little to change the situation as regards security of supply – as seen in the energy-dependency indicator, where only the Netherlands has experienced an improvement, despite standstill in its share of renewable energy. This indicates either that renewable energy use was not indigenous, or that import dependency continued to escalate for other fuels. Our conclusion remains: synergies between climate change and security-of-supply concerns have not been realized by the four countries and the EU collectively. On the other hand, the energy-dependency indicator might have been even higher without the corresponding growth in the share of renewable energy.

Table 10.5 Share of renewables in total energy consumption and net energy-import dependency, 2005/2012

	Germany	Netherlands	Norway	Poland	EU-27/EU28[1]
Share of renewables	5/10	4/4	39/44	5/9	7/11
Energy dependency	61.6/62	37.8/36.8	–609/–545 (exporter)	18.0/30.8	52.3/55.3

Notes:
[1] Croatia joined the EU in 2013.
Gross inland energy consumption in percentage/1000 tonnes. Net energy dependency in percentage of net energy imports/gross consumption. Renewables here are calculated as share of overall gross inland consumption, whereas the RED contains a specific formula for calculating the national share of renewables which leads to a higher renewable share – see Chapter 5.

Source: Eurostat (2007, 2014).

These observations indicate a complex relationship between implementation experiences and national positions. They also indicate that new circumstances have tilted member-state preferences towards a different type of policy package, to be explored below.

Towards a New 2030 Climate and Energy Framework

The German, Dutch, Norwegian and Polish positions were highly representative of three different groups of states that responded to the Commission Green Paper on the 2030 framework (Commission 2013): those wanting no unilateral GHG emissions goal, those wanting only a 40 per cent binding GHG goal, and those wanting binding goals also for renewable energy (and some for energy efficiency as well (Ibec 2013)). Poland was joined by other CEECs that wanted to await the upcoming 2015 Paris climate treaty negotiations conference and then have the EU target adjusted in line with the outcome. Others favoured 40 per cent GHG as an unconditional unilateral target, including major member-states Germany, the UK, France and Italy. The UK, the Netherlands, Norway and others wanted only this GHG target. Ministers from eight member-states, including Germany, France and Italy, joined in a letter urging the Energy and Climate Commissioners to retain also a renewable energy target (Reuters 2014a).

The European Parliament voted for three binding targets: 40 per cent GHG emissions reduction, 30 per cent share for renewables and 40 per cent higher energy efficiency. The Commission was split. Climate Commissioner Hedegaard opted for a 40 per cent GHG target but also targets for renewable energy and energy efficiency, stating that experiences with the 2020 package had shown the wisdom of the latter targets to drive restructuring after the collapse of the ETS allowance price. Energy Commissioner Oettinger held back, opting for no more than 35 per cent cut in GHGs and sharing the opinion of the Industry Commissioner that new renewable energy goals could raise energy prices and threaten the competitiveness of industries that needed to recover from the economic crisis (EurActiv 2014c).

In January 2014, the Commission adopted its compromise proposal, retaining a 40 per cent reduction of GHGs as a binding unilateral target to be achieved partly by ETS and non-ETS sectors – thus signalling a departure from international flexibility (Commission 2014e).[15] A renewable energy target of at least 27 per cent by 2030 was proposed, but at the EU level only: this was slightly above expected developments in Commission 'business-as-usual' scenarios (24 per cent) and would not be translated into new and binding national targets. No new target was proposed for energy efficiency, but the Commission stated that 25 per cent savings would be needed to reach the 40 per cent GHG reduction, and promised to analyse whether amendments to the 2012 Energy Efficiency Directive would be needed under the planned assessment in mid-2014. The Commission also proposed a comprehensive new energy governance system to compensate for setting only EU-level renewable energy targets and to enable monitoring of target achievement at the EU level – not only the renewable energy and climate targets but also security-of-supply and energy-cost parameters. The tone of the proposal had changed markedly compared with 2007: gone was the rhetoric on a new 'green' industrial revolution. CCS was not mentioned in the upscaled NER 300, and no new CCS goals were proposed.

Initial responses to the Commission's proposal showed deep divisions among the member-states. The three groups of positions on the 2013 Green Paper had now been reduced to two main groups. The first was the 'Green Growth Group', an informal grouping of like-minded energy, climate and environment ministers from 13 EU member-states, plus Norway.[16] This group issued a joint statement prior to the March 2014 European Council endorsing the core elements as set out by the Commission (Green Growth Group 2014). The other group was led by Poland, supported by other CEECs that agreed on a common list of demands.[17] Their major points were full national sovereignty over the energy mix and protection of coal, more EU subsidies to modernize energy systems, and a heavier burden on the 'rich' EU countries that argued for a more ambitious climate policy. (Basically: those who demand a more ambitious climate policy should deliver the reductions themselves.) Poland stood firm on the position that new climate and energy policies should be made conditional on the achievement of a global climate deal in Paris 2015. Compared with Poland's earlier veto of the 2050 roadmaps, the focus on demands could be said to signal a somewhat more constructive attitude.

The crisis in Ukraine that peaked in March 2014 evoked new security-of-supply concerns that also affected negotiations on the 2030 climate and energy framework. Then Polish Prime Minister Donald Tusk (later President of the European Council) responded by proposing a new EU Energy Union (EurActiv 2014d) – making the EU into a purchase cartel for gas to stand stronger vis-à-vis Russia, and with full exploitation of coal and shale-gas resources in Europe as key pillars. Tusk had repeatedly engaged in Polish energy policy, viewing nuclear, coal and shale-gas as the answer to energy dependency and expected increases in the demand for electricity, gas and oil (see also Chapter 7).

In May 2014, the Commission adopted its Communication for a new European Energy Security Strategy (Commission 2014f). This proposed strengthening of a series of long-pending proposals for contingency mechanisms that had not been properly implemented – including a new requirement for member-states to ensure 15 per cent interconnection capacity by 2030.[18] The crisis also boosted demand from the governments of some member-states, among them Germany, to include a binding energy-savings goal as part of the 2030 package in the planned review of the Energy Efficiency Directive (Reuters 2014b). The Commission followed up by proposing 30 per cent energy savings as an integrated target of the 2030 framework (EurActiv 2014a).

The adoption of climate and energy targets and policies for 2030 required agreement in the European Council, as with the targets and policies for 2020. Poland was no longer a newcomer in the EU, and one lesson from 2007/2008 stood out: know what you sign up to, for now and in the future! Poland demanded that targets and policies be negotiated together in one round, in order to get control over subsequent development of legislation based on a new framework. The position of Poland in alliance with other CEECs led to a sense of drama in the summer and early autumn of 2014. Negotiations and adoption of the package were scheduled for October. The EU had to submit its position to the international climate negotiations in Paris in the first quarter of 2015. New elections to the European Parliament were held in May and a new Commission would take office that autumn. Other major emitters were sitting on the fence, and the prospects for an ambitious and binding outcome in Paris looked bleak. In many ways, the situation resembled the negotiation circumstances in 2008, described in Chapter 5. Failure was a real possibility right up to the European Council meeting of 23–24 October 2014.

The European Council adopted the new 2030 framework – which included the new binding domestic reduction goal of at least 40 per cent reduction in GHG emissions, 27 per cent increase in renewable energy consumption (binding only at EU level) and an indicative target of 27 per cent increase in energy efficiency. Poland thus accepted new 2030 EU policies based on the ETS as the principal climate policy instrument *before* the Paris climate negotiations in December 2015. Concessions granted to Poland and its allies included the text of the first paragraph in the conclusions from the heads of states and governments, which states that the European Council will keep all the elements of the framework under review and will continue to provide strategic orientations as appropriate – notably with respect to consensus on ETS, non-ETS, interconnections and energy efficiency (European Council 2014b). This paragraph would appear to give Poland greater control over subsequent policy development. Second, the European Council would 'revert' to the framework after Paris – a formulation which could also be interpreted as a concession to Poland.

As further concessions mainly to Poland and other low-income CEEC member-states, it was agreed that burden-sharing would continue for non-ETS sectors, based on GDP/capita (as with 2020 policies). There is to be greater flexibility in how national targets can be achieved, although the details are not yet clear. For ETS sectors, free allocation will continue for energy-intensive industry, to prevent the risk of carbon leakage. Future allocation should ensure affordable energy prices and better alignment with changing production levels in different sectors – which may mean expanded opportunities for state aid to cover indirect costs.

Low-income member-states can allocate up to 40 per cent of the allowances for free to the electric power sector. This represents a weakening of current 2020 policies aimed at gradually reducing free allowances to zero by 2020. The 'solidarity fund' for modernizing energy systems in low-income member-states will continue (see Chapters 5 and 7), and upgrading of coal power is not ruled out. A new innovation fund will be based on NER 300 (now NER 400). CCS and the renewables are mentioned (as before), but the fund will be expanded to innovation in industrial sectors, including small-scale projects. Clean-coal technologies are not ruled out, but no new goals for CCS have been adopted.

Finally, the framework recognizes indigenous coal and shale gas as important for energy security options. Full respect for the freedom of member-states to determine their energy mix has been explicitly included, and the new renewable energy goal is linked to a rather vague section on (energy) governance and the idea of an energy union. Member-states had held highly differing views on the extent to which energy policy should be governed at the EU level, and on the design of such governance.[19]

In summary, the new EU 2030 framework represents a 're-packaging' compromise to satisfy the main veto players, with substantial concessions to Poland and other CEECs. The agreement also reflects varying experience with implementation – poor experience with CCS and mixed as regards the RED. The Commission's proposal clearly anticipated the positions of the member-states, and the European Council negotiations led to a number of further concessions. In essence, the reform process had a more intergovernmental flavour compared with the initiation of the 2020 targets and policy package. External factors were also important, including the economic crisis and lack of international progress on climate change. Changes in circumstances led to a new unilateral climate commitment, with reduced innovation and 'green' growth ambitions, and special arrangements for dealing with the low carbon price. However, the 2030 framework remains to be specified through detailed legislation that will determine the full ambitiousness of the new 2030 policy.

10.4 TOWARDS LOW-CARBON TRANSFORMATION?

This volume has shown that the climate and energy package for 2020 has *not* yet been a collective 'game-changer' towards 2050 low-carbon transformation. Will, then, the new 2030 framework trigger such collective transformation? We restrict the brief discussion below to the current status of new EU policies, acknowledging that transformation will be affected by further policy development and a host of other factors, such as energy technological and economic developments in the EU and the rest of the world.

Let us start with the revision of the EU ETS. This system is in crisis because of current and expected surpluses of allowances that suppress carbon prices, in turn failing to provide sufficient incentives for companies to step up investments in low-carbon solutions.[20] Through the 2030 framework, the Commission seeks to confront the problem with two countermeasures. The first is the Market Stability Reserve, aimed at creating a better balance between supply and demand of allowances and improving resilience to economic fluctuations/shocks.[21] This is a technical device that could be compared to 'automatic backloading' if the surplus is too high (see Chapter 5). It would also entail release of allowances from the Reserve, should the market be too tight. The mechanism was originally proposed to start in 2021. The European Parliament and several member-states (among them, Germany and the UK) pressed for start-up in 2017 or 2018, whereas Poland and its allies sided with the Commission proposal. On 5 May 2015, the European Parliament and the Council in trialogue meetings agreed on a compromise: the start date will be January 2019, and currently backloaded allowances will be placed directly into the Reserve.[22] Uncertainty remains as to how the Market Stability Reserve will work in practice, so the mechanism is to be reviewed within three years.[23]

The second countermeasure is to revise the ETS Directive for Phase IV, 2021–2030, with a higher cut in annual total cap of emissions (2.2 per cent as against the current 1.74 per cent). The legislative process will be guided by the MSR agreement and the conclusions in the 2030 climate and energy framework. In July 2015, the Commission proposed a revised directive, including three main elements: a more ambitious linear reduction factor (2.2 per cent) for GHG emissions, new rules for free allocation and carbon leakage, and provisions for funding innovation and modernization (Commission 2015c).

Predicting the carbon price is extremely difficult, but forecasts indicate a figure somewhat above 30 euros by 2030 with the MSR and the new cap (Fjellheim 2014).[24] To put this into perspective: back in July 2008 the carbon price reached about 30 euros, and major international oil companies are already basing their investments on a carbon price of around 35 to 50 euros (*Bloomberg Business* 2014).

The preliminary conclusion would thus be that gradual reduction of the allowance surplus may provide greater confidence in the ETS for creating stable investment signals, but carbon-price forecasts indicate that the system is unlikely to become a major driver for a low-carbon transition –

a conclusion supported by the continued emphasis on carbon-leakage provisions. CCS has been included as a technology eligible for support from the new NER 400 fund, but no new CCS targets were adopted by the heads of state in October 2014. Carbon-price forecasts for 2030 indicate prices far too low to make it feasible to revive CCS as a low-carbon solution by this date (*Bloomberg Business* 2014).

The new ESD of the 2030 framework covers the non-ETS sectors for 2021–2030. A detailed legislative proposal is planned for 2016. Current legislation on reporting and enforcement will continue; and any surpluses of AEAs under the current ESD will, in contrast to the ETS, probably be cancelled before 2021. The European Council has concluded that national targets will be based on GDP/capita and will vary between 2005 stabilization and 40 per cent emissions cut by 2030. For countries with GDP/capita above the EU average, targets will be adjusted to reflect cost-effectiveness in a 'fair and balanced manner' which is yet to be specified.

There will be no more import of external credits, but EU internal flexibility should be enhanced significantly: banking and borrowing of AEAs, transfer between member-states, and between the ESD and ETS. Auctioning of a share of AEAs is one possibility. Another, included in the European Council conclusions, is that a limited, one-off reduction of ETS allowances for ESD compliance should be made available for member-states with high abatement costs. Finally, land use, and land-use change and forestry (LULUCF) should, according to the conclusions of the European Council, be included before 2020. LULUCF, which is not part of the current climate reduction target, will reduce the EU pressure towards transformation. How much will depend on the emission accounting methodology agreed.[25]

The ESD provides a regulatory framework, but it is up to each member-state to adopt policies and measures for its non-ETS sectors. How transformative the pressure on these sectors will be is highly uncertain. On the one hand, the European Environment Agency expects emissions to decrease after 2020, but further national policies and measures will be needed (European Environment Agency 2014). On the other hand, the range of flexibility mechanisms adopted in the end may reduce the regulatory pressure on member-states.

The 2030 framework has no new specific policies for curbing emissions from transport – the top-emitting non-ETS sector. There have been major implementation problems with the Fuel Quality Directive, still not fully resolved at the EU level; and likewise as regards the Car Emissions Regulation, with watered-down emissions standards and timetables postponed. Although the ESD is likely to prove more constraining than existing policies, there is currently little to indicate any transformative changes initiated by the EU in relevant national institutions, policies or transport systems.

The new energy targets in the 2030 framework are not very ambitious – 27 per cent share for renewables in total energy consumption and 27 per cent reduction in energy compared with business-as-usual projections for 2030. The former is binding only at the EU level (no national targets), and business-as-usual scenarios indicate a 24 per cent share. Additionally, new state-aid guidelines for 2014–2020 emplace restrictions on the flexibility available to member-states in choosing support systems for renewables. Feed-in tariffs, the system most widely applied across the EU, must gradually be phased out, according to the guidelines. While the aim is to bring policies on renewables more in line with the logic of the internal energy market, it remains to be seen whether this will also curtail the rapid growth of renewables seen in countries with generous feed-in tariff systems. The experience from Germany, which revised its system in 2014 in line with the new EU guidelines, would indicate that this may result.

The energy-efficiency target is non-binding, also with targets only at the EU level. The current Energy Efficiency Directive is the main EU driver of national policies. This Directive has not been well implemented, however, with all member-states but one missing the June 2014 transposition deadline (EurActiv 2015a). Review of the Directive is scheduled for 2016; the Commission has also announced a broader energy-efficiency progress review towards 2030 in 2017 (European Commission 2014a, p. 16).

To compensate for the lack of binding national energy targets, the 2030 framework proposed a new governance framework based on national plans as well as a set of key indicators for assessing progress over time. The initial proposal here envisaged a strong and far more comprehensive integrated planning system than the scattered plans that member-states must submit under the current 2020 package. The explicit aim should be to create investor certainty, transparency and coherence, and enable EU coordination and surveillance. The proposal was met by deep mistrust from the member-states because it drew up such detailed plans for the energy mix (*Energy Post* 2014). Still, the European Council agreed on a watered-down governance system that will be developed to help ensure that the EU meets its energy-policy goals. Further details on what this will mean in practice may come when the Commission adopts a more concrete proposal.

While EU policies will probably be tighter until 2030, they are not likely to trigger any fundamental transformation of the energy system, as stated as the aim by the Commission (Commission 2015d, p. 2). However, this tentative conclusion is highly uncertain, as external conditions may change and several policies for 2030 remain to be specified.

10.5 CONCLUSIONS

Our first conclusion concerning the status of implementation of the EU climate and energy packages is that transposition of the package in national legislation has been completed in the main, after significant delays (particularly regarding CCS). Application and realization of policies and measures have varied significantly among the four countries in focus here. Poland has faced the greatest challenges; and the CCS has failed completely.

Second, we find that variation in domestic politics has proved more potent in explaining variation in implementation than 'fit' and adaptation pressure. Implementation of a package of policies has made some difference, compared with implementation of single policies. Mutual concessions when negotiating different EU policies affect the fit between EU requirements and national situations. Implementation of different policies in parallel brought increased flexibility in domestic burden-sharing and compensation opportunities.

Our third conclusion concerns the good overall correspondence between mainly poor, mixed or positive implementation experiences for national positions on new policies. Initial enthusiasm for CCS has nearly vanished, and experiences from RED implementation have been mixed among the member-states.

Our fourth conclusion is that these experiences fed into and influenced the 2030 negotiations. The outcome of the negotiations for the 2030 agreement reflected a re-packing compromise based partly on experiences with the existing targets and package for 2020 agreed unanimously in 2007/2008. It also reflected the economic crisis and lack of international progress on climate change.

We find that the compromise was largely in line with the expectations of Liberal Intergovernmentalism. The Commission became more sensitive to the concerns of the least ambitious member-states, as reflected in the closeness between the initial proposal and positions put forward in the consultations. The new 2030 climate- and energy-policy framework agreed in October 2014 included extensive concessions to Poland and other CEECs that had not experienced significant new opportunities from implementing the 2020 package. In the re-packaging that was undertaken, policies on CCS and renewables were downplayed, whereas policies on energy security (including fossil fuels) became more prominent. Further, we can note that the European Parliament, which had favoured more ambitious EU targets, has proven to have little clout. The current (2015) status of new 2030 EU targets and policies indicates that they are not likely to trigger any fundamental transformation of the energy system by 2030.

NOTES

1. The Agreement on the European Economic Area provides for a special time schedule for transposition in Norway, as explained in Chapter 9.
2. 5.73 per cent in 2013, increasing gradually to 17.56 per cent by 2020 (Commission 2015a).
3. Norway is not part of the ESD.
4. Data in this section are from Eurostat (2015a).
5. There is one recent permit for a project in the Netherlands, as well as two pending permit applications in the UK.
6. In January 2015, the Commission issued its review report of the CCS Directive, recommending revision of the Directive only after more experience had been gained with this technology in Europe (Commission 2014e).
7. The way the RED calculates the renewable energy share does not encompass energy use within the petroleum sector, thus failing to incentivize conversion from fossil fuels to renewables within a major part of overall energy consumption in Norway (see Chapters 5 and 9). This reduces the Directive's long-term impact on overall emissions reduction in Norway.
8. Similar instances of radical transformation cases in the EU, like that of Denmark, were spurred by national rather than EU policies.
9. Negotiation positions are not necessarily made explicit and they may vary significantly according to preference intensity: how important differing policies are for the negotiators. Pre-existing policies and energy-economic situation have been examined to reveal positions and preference intensity whenever needed.
10. Poland's change to a less climate change-sceptical government shortly after the adoption of the 20–20–20 targets can hardly explain the opposition and implementation challenges observed. See Chapter 7.
11. The CCS Directive may represent an exception, but our data are inconclusive. The newly elected government in October 2013 announced a re-think on Norway's CCS ambitions. However, the proposed implementing measures were submitted to a public hearing (in spring 2014) prior to the announcement of a new approach (autumn 2014), indicating that the adoption of storage provisions was not heavily dependent on the new approach.
12. The others are Belgium, Greece, the UK and Spain.
13. The ESA monitors Norwegian transposition of EU legislation that has been added to the EEA Agreement.
14. Another relevant policy area is transmission-grid extension projects, viewed as a necessary part of the *Energiewende*. Planning and permissions have been seriously delayed because of local opposition.
15. The former would have to cut emissions by 22 per cent in the period 2021 to 2030, adding to the 21 per cent target cut from 2005 to 2020.
16. This group consisted of the Netherlands, Norway and Germany, together with Belgium, Denmark, Estonia, Finland, France, Italy, Portugal, Slovenia, Spain, Sweden and the United Kingdom.

17. The Visegrad Group Countries, Romania and Bulgaria Joint Paper on the EU climate and energy framework 2020–2030. May 2014. On file with author.
18. For example, Spain still has electricity grid interconnection capacity of only 3 per cent and the Baltic states 4 per cent (EurActiv 2015b).
19. Spain and Portugal succeeded in their demand for a 15 per cent energy interconnection target by 2030. Portugal had warned that it could veto the package otherwise.
20. This point can be illustrated with two examples. In Poland, a pending decision to open a newly licensed brown-coal mine, which will take years to develop, will be affected by carbon-price expectations towards 2030 and beyond. In Germany, where importing cheap coal from the USA has proven competitive due to the low carbon price and high gas price, the future carbon price will affect the competitiveness of coal.
21. Twelve per cent of allowances are to be placed in the Reserve until the surplus is lower than 833 million allowances. The MSR will release 100 million allowances each year if the number drops below 400 million.
22. This deal followed after the Czech Republic defected from the Poland-led blocking minority alliance (*Carbon Pulse* 2015).
23. Assessments, hedged with high uncertainty, indicate that oversupply will fall below the 833 million threshold in 2024 and that no allowances will be transferred to the MSR from 2025.
24. Based on forecasts by Thomson Reuters Point Carbon.
25. A proposal from the Commission would not come until 2016, that is, after the Paris Summit. NGOs have noted that inclusion of LULUCF could remove five percentage points from the 40 per cent target for 2030 (EurActiv 2015c).

REFERENCES

Bloomberg Business (2014), 'CO2 price to spur carbon capture will fall from $125, Shell says', 29 May, accessed 18 June 2015 at http://www.bloomberg.com/news/articles/2014-05-29/co2-price-to-spur-carbon-capture-will-fall-from-125-shell-says.

Carbon Pulse (2015), 'EU nations agree to 2019 MSR start date after Czechs defect', 29 April, accessed 5 June 2015 at http://carbon-pulse.com/eu-nations-agree-to-2019-msr-start-date-french-envoy/.

Commission (2013), *Green Paper: A 2030 Framework for Climate and Energy Policies*, COM (2013) 169 final, 27 March, Brussels: European Commission.

Commission (2014a), *Energy Efficiency and its Contribution to Energy Security and the 2030 Framework for Climate and Energy Policy*, COM (2014) 520 final, 23 July, Brussels: European Commission.

Commission (2014b), 'Support to the review of Directive 2009/31/EC on the geological storage of carbon dioxide (CCS Directive)', accessed 30 April 2015 at http://www.ccs-directive-evaluation.eu/assets/CCS-Directive-evaluation-Final-Report.pdf.

Commission (2014c), *Progress towards Achieving the Kyoto and 2020 Objectives*, COM (2014) 689 final, 28 October, Brussels: European Commission.

Commission (2014d), *Proposal for a Decision of the European Parliament and of the Council concerning the establishment and operation of a market stability reserve for the Union's greenhouse gas emission trading scheme and amending Directive 2003/87/EC*, COM (2014) 20/2, accessed 15 October 2015 at http://ec.europa.eu/clima/policies/ets/reform/docs/com_2014_20_en.pdf.

Commission (2014e), *A Policy Framework for Climate and Energy in the Period from 2020 to 2030*, 22 January, COM(2014) 15 final, Brussels: European Commission.
Commission (2014f), *European Energy Security Strategy*, COM (2014) 330 final, Brussels: European Commission.
Commission (2015a), *Free Allocation based on Benchmarks*, updated 8 May, accessed 30 April 2015 at http://ec.europa.eu/clima/policies/ets/cap/allocation/index_en.htm.
Commission (2015b), *Reducing Emissions from Transport*, updated 8 May, accessed 30 April 2015 at http://ec.europa.eu/clima/policies/transport/index_en.htm.
Commission (2015c), *Proposal for a Directive of the European Parliament and of the Council amending Directive 2003/87/EC to enhance cost-effective emission reductions and low carbon investments*, COM (2015) 337 final, 15 July, Brussels: European Commission.
Commission (2015d), *Energy Union Package, a Framework Strategy for a Resilient Energy Union with a Forward-Looking Climate Change Policy*, COM (2015) 80 final, 25 February, Brussels: European Commission.
Dupont, C. and S. Oberthür (2015), 'Conclusions: lessoned learned', in C. Dupont and S. Oberthür (eds), *Decarbonizing in the European Union – Internal Policies and External Strategies*, Houndmills and New York: Palgrave Macmillan.
Energy Agreement (2013), *English Summary of Energieakkord voor duurzame groei*, The Hague: Social and Economic Council of the Netherlands (SER), accessed 15 October 2015 at http://www.energieakkoordser.nl/energieakkoord.aspx.
Energy Post (2014), 'EU deeply divided over 2030 climate and energy policy', 5 March, accessed 18 June 2015 at http://www.energypost.eu/eu-divided-2030-climate-energy-policy/.
EurActiv (2014a), 'Brussels puts 30 per cent energy savings target on the table for 2030', 24 July, accessed 15 January 2015 at http://www.euractiv.com/sections/energy/brussels-puts-30-energy-savings-target-table-2030-303679.
EurActiv (2014b), 'Member states' positions on 2030 climate and energy targets revealed', 23 October, accessed 4 November 2014 at http://www.euractiv.com/sections/energy/member-states-positions-2030-climate-and-energy-targets-revealed-309279.
EurActiv (2014c), 'Oettinger rallies opposition to 2030 CO2 target', 29 January, accessed 5 June 2015 at http://www.euractiv.com/energy/oettinger-rallies-opposition-203-news-533084.
EurActiv (2014d), 'Poland calls for EU energy union', 2 April, accessed 9 April 2014 at http://www.euractiv.com/sections/energy/poland-calls-eu-energy-union-301303.
EurActiv (2015a), 'Member states face legal action over energy efficiency', 26 March, accessed 12 October 2015 at http://www.euractiv.com/sections/energy/member-states-face-legal-action-over-energy-efficiency-313259.
EurActiv (2015b), 'Energy union aims for elusive 10 per cent power grid interlinkage', 25 February, accessed 2 March 2015 at http://www.euractiv.com/sections/energy/energy-union-aims-elusive-10-power-grid-interlinkage-312383.

EurActiv (2015c), 'Europe quietly shelves forestry debate ahead of Paris climate talks', 12 May, accessed 5 June 2015 at http://www.euractiv.com/sections/climate-environment/europe-quietly-shelves-forestry-debate-ahead-paris-climate-talks-314486.

European Council (2014a), *Communication from the Commission on 'A policy framework for climate and energy in the period from 2020 to 2030', replies received from the Czech Republic and the Netherlands to the Presidency questions (11 June)*, Brussels: General Secretariat of the Council.

European Council (2014b), *European Council (23 and 24 October 2014) Conclusions on 2030 Climate and Energy Policy Framework*, SN 79/14, Brussels: General Secretariat of the Council.

European Environment Agency (2014), *Trends and Projections in Europe 2014: Tracking Progress towards Europe's Climate and Energy Targets for 2020*, European Environment Agency Report No. 6, Copenhagen: EEA.

Eurostat (undated), *Share of Renewable Energy Fuel Consumption of Transport*, accessed 12 June 2015 at http://ec.europa.eu/eurostat/tgm/table.do?tab=table&init=1&language=en&pcode=tsdcc340&plugin=1.

Eurostat (2007), *Energy: Yearly Statistics 2005*, accessed 30 April 2015 at http://ec.europa.eu/eurostat/documents/3217494/5613164/KS-PC-07-001-EN.PDF/a9aa88dc-59c7-40fc-9081-db5f7fb44813?version=1.0.

Eurostat (2014), *Energy Balance Sheets, 2011–2012*, Eurostat Statistical Books, accessed 30 April 2015 at http://ec.europa.eu/eurostat/documents/3217494/5785109/KS-EN-14-001-EN.PDF/16c0ac97-7dd6-4694-b22d-e77a36cb4e86?version=1.0.

Eurostat (2015a), *Share of Renewables in Energy Consumption up to 15 Per Cent in the EU in 2013*, 43/2015, 10 March, accessed 30 April 2015 at http://ec.europa.eu/eurostat/documents/2995521/6734513/8-10032015-AP-EN.pdf/3a8c018d-3d9f-4f1d-95ad-832ed3a20a6b.

Eurostat (2015b), *Sustainable Development – Transport*, last modified 19 February 2015, accessed 3 June 2015 at http://ec.europa.eu/eurostat/statistics-explained/index.php/Sustainable_development_-_transport.

Eurostat (2015c), *Europe 2020 Indicators – Climate Change and Energy*, last updated 5 March 2015, accessed 15 May 2015 at http://ec.europa.eu/eurostat/statistics-explained/index.php/Europe_2020_indicators_-_climate_change_and_energy.

Fjellheim, H. (2014), 'Reform av kvotesystemet: EUs viktigste klimapolitiske veivalg', *Energi og Klima*, 8 October.

German Energy Agency (2014), *Germany in 2050 – a Greenhouse Gas-Neutral Country*, Climate change 07/2014, Dessau-Rosslau: Umweltbundesamt.

Green Growth Group (2014), 'Green Growth Group Ministers' statement on climate and energy framework for 2030, joint statement on 3 March 2014', accessed 2 February 2015 at www.gov.uk/government/news/green-growth-group-ministers-statement-on-climate-and-energy-framework-for-2030.

Ibec (2013), *Summary of Member State Responses to the 2030 Green Paper for a Climate and Energy Framework*, accessed 14 April 2015 at http://www.ibec.ie/IBEC/DFB.nsf/vPages/Energy~Resources~summary-of-member-state-responses-to-the-2030-framework-on-14-10-2013/$file/Member+State+responses+to+2030+green+paper.pdf.

Ministry of the Environment [Poland] (2013), 'Communication from the Commission – Green Paper – A 2030 framework for climate and energy policies. Responses by the Government of Poland within public consultations', Warsaw, accesses 30 April 2015 at http://ec.europa.eu/energy/en/consultations/consultation-climate-and-energy-policies-until-2030.

Norway (2014), *Norwegian Position on the Proposed EU Framework for Climate and Energy Policies towards 2030*, 6 October, accessed 15 October 2015 at http://www.regjeringen.no/upload/UD/Vedlegg/Protokoll/141006-posisjonspapir-EU-klima-energi.pdf.

Piria, R., A. Arcipowska, C. Bausch, P. Hockenos, S. Müller-Kraenner and J. Ondřich (2014), *Greening the Heartlands of Coal in Europe: Insights from a Czech–German–Polish Dialogue on Energy Issues*, Prague/Berlin: Heinrich-Böll-Stiftung, accessed 18 June 2015 at https://www.boell.de/sites/default/files/greening_the_heartlands_of_coal_in_europe.pdf.

Reuters (2014a), 'Group of ministers urges EU to set 2030 goal for renewable energy', 6 January, accessed 3 June 2015 at http://uk.reuters.com/article/2014/01/06/eu-renewables-climate-idUKL6N0KG1Z720140106.

Reuters (2014b), 'Ministers from 7 EU nations call for binding energy saving goal', 17 June, accessed 5 June 2015 at http://uk.reuters.com/article/2014/06/17/eu-energy-efficiency-idUKL5N0OY4KQ20140617.

Skjærseth, J.B. (2014), *Implementing EU Climate and Energy Policies in Poland: From Europeanization to Polonization?*, FNI Report 8/2014, Lysaker: Fridtjof Nansen Institute.

Skjærseth, J.B. and P.O. Eikeland (eds) (2013), *Corporate Responses to EU Emissions Trading*, Aldershot: Ashgate.

11. Conclusions and the road ahead

Jon Birger Skjærseth and Per Ove Eikeland

The EU has adopted increasingly ambitious climate and energy targets and policies for decarbonizing the economy by 2050. How have the various member-states managed to agree? Have the policies been implemented domestically, triggering a transformation towards low-carbon economies? What are the consequences of these implementation experiences for the adoption of new EU policies to 2030 and beyond? The answers to these questions may help decision-makers to respond to the threat of climate change. EU climate and energy policy is an ideal case for learning, as the field has seen significant development over time, with successes as well as failures.

In addition to asking empirical questions, in this book we have developed novel analytical tools for explaining why, how and with what consequences EU policies have developed. Our starting point was that the development of EU climate policy has departed fundamentally from the positions of the 'least ambitious' actors also when unanimity has been required. EU climate policy comes close to the positions of the 'most ambitious' actors. Negotiation theories of issue-linkage show that agreement on ambitious policies under unanimity conditions can be reached by linking issues and simultaneously settling policies in packages that create synergies, offer compensation, and combine different interests and values.

The chapters in this volume have explored the opportunities and pitfalls of linking policies in packages, studying the full range of policy phases, from policy initiation and decision-making, through to implementation and reform. The main analytical contribution of this book lies in its systematic combination of negotiation theory with theories on EU policymaking and implementation, offering insights into how climate ambitions can be raised by linking energy and climate policies in different policy phases. The analytical framework guiding the empirical analysis was set out in Chapter 2. In this final chapter, we summarize the empirical findings of Chapters 3 to 10, extract analytical implications in relation to Chapter 2, and present some thoughts about the future road for EU climate and energy policy.

11.1 EMPIRICAL FINDINGS OF THE INDIVIDUAL CHAPTERS

Chapter 3 analysed the evolution of EU climate and energy policies from the early 1990s until the mid-2000s, to provide a background and baseline for explaining subsequent changes towards the 2020 climate- and energy-policy package. Early in this period, the EU attempted to craft a climate and energy package of policies intended to show leadership-by-example at the 1992 Rio Conference. This noteworthy but ultimately unsuccessful attempt was based on the recognition of functional links between climate and energy policies. Both policy fields would have to be adapted so as not to distort the internal energy market, with a harmonized CO_2/energy tax proposed as overarching policy instrument. However, neither the Commission nor the member-states managed to exploit these linkages politically so as to overcome distributional challenges and promote synergies. In the remaining period leading up to the 1997 Kyoto Protocol, the development of EU climate and energy policy stagnated. Ambitions of a harmonized EU-level CO_2/energy tax faded as a result of EU unanimity requirements and continued opposition among member-states and energy-intensive industry.

After the adoption of the Kyoto Protocol, climate and energy policies gained momentum in isolation, based on different timing, policy instruments, organization and interests within the Commission. In order to deliver on the EU's Kyoto Protocol reduction commitment and show 'leadership by example' aimed at convincing other countries to ratify the Protocol, an emissions-trading system was adopted. This new instrument shared features with the carbon/energy tax in conforming with the internal market logic of providing price signals to the market, and let market actors choose options for emission cuts. Unlike the tax, however, the EU ETS could be adopted by qualified majority voting. The energy policy focus changed in response to rapidly increasing oil and gas prices that affected EU industrial competitiveness and the growing import dependency that made the EU more vulnerable to supply disruptions. The trade-off between mitigation of climate change and security of supply became increasingly evident in internal EU discussions, with fears that high commitments to the former would rule out the continued use of indigenous fossil-fuel resources, mainly coal. New energy legislation was adopted to promote renewable energy and energy efficiency. These policies combined climate and security-of-supply concerns, but lacked sufficient scope and ambition. Targets were non-binding on the member-states, which were keen to protect their specific energy-economic interests.

From late 2003, compartmentalization within the Commission developed into open conflict on the costs of climate policies. Different Commission services catered to different stakeholders with highly differing interests and positions on climate and energy policies. This conflict impeded the initiation of more ambitious long-term EU climate targets. From Rio to Kyoto and beyond, then, neither the member-states nor the Commission were able to exploit functional links between climate and energy issues politically in a way to move beyond the 'least ambitious' actors.

Chapter 4 assessed the initiation of the EU climate and energy package for 2020. How did the EU manage to combine the detached and conflictual issues in climate and energy policy? First, different policies were either linked or decoupled to provide common grounds for agreement. This ensured that differing priorities among pivotal individual and collective decision-makers as regards energy security and climate mitigation were successfully aligned. Linking policies in a package that combined interests and values and promised synergies rather than trade-offs between objectives helped to unify support for more ambitious EU climate and energy policies within the Commission and among the member-states. Linking CCS to the policy package proved politically decisive: the trade-off between fossil-fuel-based energy security and climate concerns could be broken, reducing opposition from powerful fossil-fuel interests – including oil companies and coal-based member-states that had grown in number with the accession of the CEECs.

Other functionally related policies were either deliberately de-linked or were simply not included in the emerging policy package. Including energy efficiency in the package could have increased complexity and exacerbated various conflicts of interest, particularly between the CEECs and the EU-15. Also climate policies on road transport – the Fuel Quality Directive and the Car Emissions Regulation – were de-linked from the emerging package, as they would serve to increase the regulatory burden for only a few member-states that had car manufacturers and petroleum producers. Including these transport policies might serve to split rather than unify the interests and preferences of the member-states. Finally, the drafting of the Renewable Energy Directive failed to include a harmonized EU-level policy instrument that would also promote progress in realizing the internal energy market. Harmonization of renewable energy policy in line with the internal energy market logic would have split the member-states and destroyed the common ground for the climate and energy policy package.

Second, the Commission promised even broader policy synergies and complementarities. The focus on trade-offs was replaced by a heavy emphasis on complementarities and synergies between climate and energy policies, and between these policies and other policies. Synergies were to be created by mutually reinforcing climate and energy goals, so as to reduce air pollution and energy dependency, create 'green' jobs and stimulate energy-technological innovation. However, potential conflicts with other functionally related issues were largely swept under the carpet, such as the internal energy market's downward pressure on energy prices that would lower incentives for alternative energy investments.

Both the Multi-level Governance and the Liberal Intergovernmentalist approaches can help to explain why the package was initiated. In line with LI, the EU member-states collectively and individually changed the Commission's room for manoeuvre to initiate new energy and climate policies. Further, member-states put pressure on the Commission, which was initially unable to act forcefully due to internal rivalry. No single member-state can be said to have been an 'entrepreneurial leader' in initiating the package. The member-states invited rather than requested the Commission to initiate new energy and climate policies and did not request any specific package design.

In line with MLG expectations, it was the Commission, and specifically DG Environment, that took the initiative to the package in 2006 and allied with DGTREN. After a history of conflicts between these two DGs, a new Commission with changes in key personnel and coordinating procedures facilitated joint efforts. Also external energy events and international commitments played a role in elevating climate- and energy-security issues on the EU political agenda. Importantly, the Kyoto Protocol was ratified in 2005, ending the 'what if …' discussion and neutralizing those who had been voicing concerns about the costs if the Protocol should fail to gain sufficient ratifications for it to enter into force.

Chapter 5 examined how new and binding EU policies were negotiated, agreed and eventually linked or combined in 2008 to deliver on the '20–20–20' targets. The Commission proposed a 'core' package consisting of two cross-sectoral policy instruments: a revised EU ETS and an ESD for the non-ETS sectors, and two low-carbon specific Directives, on renewable energy sources and CCS. The combination of policies in the final outcome of the negotiations was mainly in line with our expectations from issue-linkage theories. Issues and policies that yielded opportunities for joint gains for pivotal decision-makers were linked in the core package; other functionally related policies were excluded from joint negotiations. This made possible agreement to joint policies exceeding the positions of the 'least ambitious' actors.

The proposed package also provided ample room for mutual concessions and compensation for less well-off member-states. The room for mutual concessions across linked policies was exploited by the pivotal actors in the negotiations, the member-states and the European Parliament. The initial design of the package also strengthened the cooperative nature of the negotiations. The package was based on common impact assessment of the various package components aimed at cost-sharing according to GDP/capita. Compensation by side-payments was ensured by redistributing revenues from auctioning of emissions-trading allowances.

External factors played a role also in this phase. The Kyoto Protocol became important for reducing expected EU costs through opportunities for the purchase of external credits for compliance with the ETS, the ESD and also the Fuel Quality Directive. The international climate negotiations on a post-Kyoto treaty facilitated the package by serving as a target for EU leadership-by-example ambitions and by emplacing a time-pressure on the 2008 negotiations. However, lack of progress on a new treaty and the financial crisis led to demands for renegotiating the package. These were fended off, however, on the grounds that new negotiations might threaten the fragile political consensus already achieved. This shows one pitfall in making complex package deals – it can be hard to adapt to new circumstances.

Our main conclusion here is that negotiation theory on issue-linkages can explain *how* the interests and values of pivotal policymakers were combined and agreement reached by the initiation and negotiation on a 'core' policy package and the de-linking of other policies that might split interests or make the negotiations too complex. LI, MLG and external factors can explain *why* different policies developed in the first place. As EU policymaking on energy has traditionally been more intergovernmental in character than climate policy, the combined energy and climate package can be explained by Liberal Intergovernmentalism and Multi-level Governance in combination.

Chapters 6 to 9 analysed the implementation of the agreed EU climate and energy package in Germany, the Netherlands, Norway and Poland, with the results drawn together in a comparative analysis in Chapter 10. A first conclusion is that high hopes in Brussels have partly been dashed by the member-states. Transposition of the package in national legislation has been completed after significant delays, particularly as regards the

CCS Directive. Application of policies and measures has varied considerably among the four countries studied here. Poland has encountered the greatest challenges, but CCS has failed completely in all member-states. No new CCS pilot plants have been constructed or are likely to be constructed by 2020, despite the 2007 European Council agreement that provides for up to 12 CCS pilot plants by 2015. The massive cut in GHG emissions so far in the implementation phase has been caused partly by the economic crisis, which masks deficiencies in adopting new policies. The package has not yet served as a collective 'game-changer' placing the countries on pathways towards low-carbon economies. The only partial exception – Germany's radical transformation towards renewable electricity production – had been started long before the EU climate and energy package was adopted, and was further spurred by the Fukushima nuclear disaster in Japan.

The second conclusion is that variation in domestic politics can explain more of the variation in policy implementation than can 'fit' and adaptation pressure. Opposition (as well as support) from societal actors affected by EU policies and fragmented administrative organization proved important explanations across the cases, whereas change in government and policy style proved less important. That being said, however, varying combinations of factors can explain implementation patterns in the countries studied. This underscores how domestic political factors interact differently in varying national contexts.

The fact that implementation concerned an agreed package of policies – not a single directive – has made some difference. Mutual concessions given when negotiating different policies did affect fit and adaptation pressure, specifically for Poland. Implementation of different policies in parallel has provided some greater flexibility in domestic burden-sharing and compensation opportunities, mainly facilitating implementation. However, the four countries have only to a limited extent been able to translate EU-level envisaged benefits to the member-state level, including keeping coal carbon-neutral through CCS, 'green' jobs and energy-technological innovation, with variation between them, however.

The third conclusion concerns the strong overall correspondence between implementation experiences and positions on new policies – particularly for CCS and renewables. Mixed experiences with different policies in the package were fed into and influenced the reform of 2020 policies and targets for 2030. The four countries in focus here proved representative of the main groups that formed in the negotiations on reform for 2030.

And our fourth conclusion: the new 2030 climate and energy policy framework agreed in October 2014 represents a re-packaging of existing policies: CCS and renewables are downplayed whereas the ETS and ESD will be strengthened. Two observations indicate that the reform phase has become more intergovernmental in nature and in line with LI expectations. First, the Commission has become less able to initiate new ambitious policies that build consensus for a common EU project towards decarbonization. This is illustrated by failed efforts on the long-term climate and energy roadmaps that were blocked by Poland. Second, the increased opposition from Poland and its allies significantly shaped the 2030 climate and energy framework outcome. The new 2030 framework includes several concessions to Poland and other CEECs that had not experienced significant benefits or synergies with other national goals.

We have seen how the initiation, adoption, implementation and reform phases of EU climate and energy policies have become interrelated. Policy-linking within one phase has affected the outcomes in the next phase. The initiation of new linked policies has combined interests and values among pivotal actors and provided synergies. This has facilitated cooperative negotiations and unanimous adoption of new policies that have provided wide scope for mutual concessions and side-payments to compensate low-income member-states. Overall, this has led to a good fit between EU policy requirements and the national situations in our four case-study countries. Implementation, however, partly failed as a result of domestic politics challenges. The EU package approach has proved inadequate for translating EU-level benefits to the member-states. As a result, negative and mixed experiences from implementation led to re-packing of new targets and policies for 2030.

11.2 ANALYTICAL IMPLICATIONS OF THE STUDY

Our first conclusion here concerns the initiation and negotiation of climate and energy policies. The combination of negotiation theory and traditional theories for explaining EU integration and policymaking has proven essential for understanding how and why new EU policies have become more ambitious under conditions of unanimity. Pivotal actors had differing interests and objectives: some considered climate-change mitigation as most important; others focused on security of supply and competitiveness as top priorities for policy development. The package managed to combine policies that underscored the opportunities for synergies rather than trade-offs: *all* the objectives could be realized. Further, our study has found it useful to differentiate between conditions that promote package solutions and package trouble. The issue-linkage propositions set out in Chapter 2 have proven particularly useful for clarifying why some policies were linked in a package and others not. MLG and LI approaches in combination are well suited for explaining why EU climate and energy policies developed in the first place.

Our second analytical conclusion concerns implementation and reform. The approach focused on ‘fit’ and adaptation pressure between EU requirements and the national status quo. This approach has been modified to address shortcomings and is still regarded as a potentially powerful and parsimonious explanation within the Europeanization and implementation literature. However, ‘fit’ has shown limited explanatory power as regards variation in policy implementation. Here the alternative domestic politics perspective seems preferable: variation in governments, how societal actors were affected, policy styles and administrative organization can explain much of the variation in implementation. An intriguing observation is how differences in access to domestic decision-making for renewable energy interests between Poland and Germany have contributed to produce different ambitions and support systems for renewables in electricity production. Our study has also shown that understanding how new EU policies are negotiated and linked is important for explaining policy implementation in the member-states. Finally, how EU policies are implemented has consequences for national positions on new policies to respond to long-term challenges. This means that the study of domestic implementation experiences is essential for explaining reform of climate policies for the longer term.

Our third analytical conclusion concerns the study of the policy cycle – from initiation to reform of policies. Since the outcomes of each phase affect the next phase, negotiation theories and theories of EU policy-making and implementation should be applied across policy phases. A policy-cycle approach can show how issue-linkages and policy packages may serve different functions in different phases that may advance or impede subsequent policy progress – as well as how and why the EU policy dynamic between member-states, EU institutions and factors external to the EU might change over time.

11.3 THE ROAD AHEAD

The EU will need to reform and tighten the 2030 climate and energy framework in several policy rounds in order to deliver on its 2050 decarbonization goal. The challenge is formidable: to facilitate and shape a transformation towards decarbonization among 28 different member-states by the year 2050.

The preceding conclusions show that at least three conditions are necessary. First, consensual EU-level policies can progress only by linking issues and policies in a way that combines different interests and values among pivotal actors. Second, domestic implementation will serve as feedback facilitating further steps, to the extent that it can generate benefits and positive experiences. Third, key emitting EU partners and competitors must agree on comparable commitments within the international climate cooperation to level the playing field. We will briefly discuss these conditions in light of the first policy signals by the new Juncker Commission and international events.

The Energy Union has been made a central priority of the Juncker Commission, with a new vice-president tasked with improving horizontal coordination to enable realization of this venture. Moreover, a Commissioner has been appointed for joint climate *and* energy policy execution. The Commission's first Communication on the Energy Union Package was framed as a *coherent* climate and energy package (Commission 2015a). However, the Communication reflects the deep divisions among member-states, as also expressed during the European Council negotiations in October 2014.[1] Concerning the link between climate and energy policies, the text is ambiguous on how to deal with fossil fuels in light of the new focus on energy security. On the one hand, the Communication states that EU goals necessitate moving 'away from an economy driven by fossil fuels'. On the other hand, it signals that also conventional and non-conventional fossil fuels may be an option for increasing the share of domestically produced energy. The objective of decarbonizing the economy has been retained, but the specific emissions reduction target for 2050 is not made explicit.[2] The climate objective is not accorded the position of primary driver of energy policy development that it had back in 2007.

Policymakers in Brussels seem undecided as to how to pursue policies that can combine different interests, after CCS failed to bridge climate-change and energy-security concerns. As it stands now, the Energy Union initiative reflects a more diversified and re-nationalized landscape compared with the 2020 package. In the absence of CCS that can give the fossil-fuel industry a future in decarbonization, policies promoting equivalent technologies need to be linked for political reasons, to reduce opposition and promote EU-level policies further.

Another priority in the Energy Union initiative is to improve the *implementation* of current legislation particularly to complete the internal energy market. A well-functioning single energy market is seen as the main instrument for achieving energy-security goals, with harmonization of national policies and regulations as fundamental for realizing the single market. This may create positive experiences for the public and industry by reducing energy prices, but low energy prices also counteract investments in energy efficiency and renewables. Concerning the climate and energy package, implementation experiences with CCS, the ESD, ETS and RED have been mixed among member-states. This is reflected in the new 2030 framework, which strengthens the climate target based on the ETS and ESD, weakens EU targets on renewables and nearly stops CCS. As a result of the economic crisis, among other factors, the ETS and ESD have not worked as intended. The increased flexibility of these instruments agreed in the 2030 framework fit the varying needs of the member-states, which differ widely in energy-economic and climate-policy interests and preferences. Still, a key challenge is to minimize costs and promote continuous benefits when these policies will have to be strengthened and implemented by 2030, 2040 and 2050.

That in turn will depend partly on how the international climate negotiations develop. Disagreement within the EU has repercussions also for the EU's leadership ambitions and its position at the international climate negotiations in Paris compared with Copenhagen 2009.[3] The EU leaders concluded in their 2030 climate and energy framework that they would 'revert to the issue' after the international climate negotiations in Paris in December 2015. Theoretically, this could mean that the EU will increase its ambitions after Paris if there is a new treaty that appears sufficiently ambitious. However, it seems unlikely that the EU will strengthen its 2030 climate and energy targets as a result of a Paris treaty. That might also mean that the EU could lower its ambitions if the Paris meeting proves to be a failure. However, also that seems unlikely, since the targets included in the EU 2030 climate and energy policy framework are 'domestic' ones. Still, high ambitions in Paris could strengthen the development of new EU legislation. Beyond 2030, it will be difficult for the EU to go it alone.

How the future will look is veiled in great uncertainty. On the one hand, institutionalized cooperation may gather momentum through a 'snowball effect' generating positive feedback from implementation and facilitating further steps. On the other hand, it might be more in line with the economists' 'law of diminishing returns'. Here, the first steps are likely to be the 'easy ones' in which marginal benefits clearly exceed marginal costs. This 'law' indicates that it would become increasingly difficult to promote new joint policies, and governments would gradually become more reluctant as regards implementing them.

These different dynamics hinge on several other factors. Will the European Commission regain its entrepreneurial climate-policy role in the new diversified and renationalized landscape? Will Poland and other

CEECs experience benefits from transforming their energy systems? Will Germany and other frontrunners slow down their investments in renewable energy? And will the world leaders manage to agree on and implement an ambitious climate treaty that can enable the EU to maintain and strengthen its climate ambitions?

As the EU's long-term plans stand now, the decarbonization challenge lies increasingly with implementation in the member-states. To ensure longer-term collective success, climate, energy and possibly other policies will have to be regionalized or linked through repeating policy cycles in ways that can create positive experiences and benefits among all the diverse member-states of the European Union.

NOTES

1. The Energy Union initiative was triggered by the Russia–Ukraine crisis and picked up by Donald Tusk, who set energy security as the top priority in new climate and energy policy development when he served as Poland's Prime Minister – later becoming President of the European Council.
2. The only reference to 2050 is the call for a new and critical forward-looking approach to carbon capture and storage and carbon capture and use in the power and industrial sectors.
3. The EU position is: 'a vision for a transparent and dynamic legally binding agreement, containing fair and ambitious commitments from all Parties based on evolving global economic and geopolitical circumstances' (Commission 2015b).

REFERENCES

Commission (2015a), *Energy Union Package, A Framework Strategy for a Resilient Energy Union with a Forward-Looking Climate Change Policy*, COM (2015) 80 final, 25 February, Brussels: European Commission.

Commission (2015b), *Energy Union Package, The Paris Protocol – A Blueprint for Tackling Global Climate Change beyond 2020*, COM (2015) 81 final, 25 February, Brussels: European Commission.

Interviews

CHAPTERS 3–5

- Chris Beddoes, Europia, Vice-Secretary General, 15 September 2011
- Arno Behrens, Centre for European Policy Studies, Head of Energy & Research Fellow, 24 April 2011
- Simon Bennett, DG ENER, Policy officer CCS Policy, 23 April 2012
- Sylvia Beyer, DG ENER, Policy Officer Unit Security of Supply and Networks, 23 April 2012
- Tom Brookes, European Climate Foundation, Head of Energy Strategy Centre, telephone interview, 9 May 2012
- Helen Donoghue, DG ENER, Principal Administrator of Strategy and Programming Unit, 25 April 2012
- Martina Dopplehammer, Unit Low Carbon Technologies, DG CLIMA, 25 April 2012
- Christian Egenhofer, Centre for European Policy Studies, Senior Fellow, Head of the Energy and Climate programme, 13 September 2011 and 24 April 2012
- Paal Frisvold, Bellona Brussels, Board chairperson, 11 September 2011
- Paul Greening, Director Emissions & Fuel, ACEA, 25 April 2012
- Bjørn Staale Haavik, Energy Advisor, Norway's Representation to the European Union, 14 September 2011
- Ian Hodgson, Policy Officer Transport and Ozone, DG CLIMA, 24 April 2012
- Tor Eigil Hodne, Statnett Brussels, Director EU Office, 14 September 2011
- Paul Hodson, DG TREN, Head of Unit Energy Efficiency, 13 September 2011
- Terhi Lehtonen, Environment Advisor Greens/EFA Group, European Parliament, 24 April 2012
- Jürgen Müller, DG CLIMA, Member of Cabinet, 24 April 2012

- Susanne Nies, Eurelectric, Head of Unit Energy Policy & Generation, 23 April 2012
- Philip Owen, Head of Unit Transport and Ozone, DG CLIMA, 26 April 2012
- Brigitta Renner-Loquentz, DG COMP, Head of Unit State Aid, 26 April 2012
- Jürgen Salay, DG CLIMA, Policy Officer, coordinator of the implementation of the Effort-Sharing Decision, Unit Low Carbon Technologies, 13 September 2011
- Raphael Sauter, Unit Low Carbon Technologies, DG CLIMA, 25 April 2012
- Jesse Scott, Eurelectric, Head of Unit Environment & Sustainable Development Policy, 26 April 2012
- Yvon Slingenberg, DG CLIMA, Head of Unit Implementation of ETS, 12 September 2011 and 26 April 2012
- Matti Supponen, DG ENER, Policy Coordinator, policy and project officer of Unit Internal Market II: Wholesale markets; electricity and gas, 13 September 2011
- Jonas Teusch, Researcher, Centre for European Policy Studies, 24 April 2011
- Kai Tullius, DG TREN, Policy Coordinator CCS Policy, 23 April 2012
- Tom van Ierland, DG CLIMA, Policy Officer, Economic assessment of climate policies in Unit Strategy and Economic Assessment, 14 September 2011
- Hans van Steen, DG ENER, Head of Unit Renewables and CCS Policy, 25 April 2012
- Ignacio Vazquez Larruscain, DG CLIMA, 26 April 2012
- Stefaan Vergote, DG CLIMA, Head of Unit Strategy and Economic Assessment, 14 September 2011
- Peter Vis, DG CLIMA, Head of Cabinet, 15 September 2011
- Torsten Wöllert, Unit Low Carbon Technologies, DG CLIMA, 25 April 2012
- Thomas Wyns, CAN Europe, Policy Officer, 15 September 2011

CHAPTER 6

- Marius Backhaus, Federal Ministry for Economic Affairs and Energy, 6 October 2014, Berlin
- Jasper Braam, Federal Ministry of the Environment, 5 June 2014, Berlin

- Bjørn Erik Brustad, Norwegian Ministry of Foreign Affairs, 21 May 2014, Oslo
- Ernst Peter Fischer, Deputy Director-General, Federal Foreign Office, 6 October 2014, Berlin
- Severin Fischer, Dipl. Pol., Stiftung Wissenschaft und Politik (SWP), 4 June 2014, Berlin
- Sabine Gores, Öko-Institut e.V.m, telephone interview, 2 June 2014
- Joachim Hein, BDI, 6 June 2014, Berlin
- Rainer Hinrichs-Rahlwes, member of Board, German Renewable Energy Industry Association, 6 June 2014, Berlin
- Dr Volker Hoppenbrock, M.A., Federal Ministry of Economic Affairs and Energy, 5 June 2014, Berlin
- Dr R. Andreas Kraemer, Director, Ecologic Institute, 5 October 2014, Berlin
- Dr Joachim Lang, Director E.ON, 6 October 2014, Berlin
- Dr Felix Matthes, Öko-Institut, 4 June 2014, Berlin
- Henning Mümmler-Grunow, German Bundestag Office of Oliver Krischer, 4 June 2014, Berlin
- NN, Federal Ministry for Economic Affairs and Energy, 3 June 2014, Berlin
- Dimitri Pescia, Senior Project Manager, Agora Energiewende, Berlin 7 October 2014
- Christian Redl, Senior Associate, Agora Energiewende, 5 June 2014, Berlin
- Justus Riedlinger, Ministry of Energy, Agriculture, the Environment and Rural Areas, Schleswig-Holstein (telephone interview, 30 May 2014)
- Stephanie Ropenus, Agora Energiewende, 5 June 2014, Berlin
- Dr Martin Ruhrberg, Der Bundesverband der Energie- und Wasserwirtschaft e. V. (BDEW) 6 June 2014, Berlin
- Dr Jan Scharlau, Federal Ministry of the Environment, 4 June 2014, Berlin
- Einar Westre, Execute Director, Energy Norway (telephone interview, May 2014)

CHAPTER 7

- Marta Babicz, Head of European and Regional Energy Policy Unit, Ministry of Foreign Affairs
- Maciej Bukowski, Director, Warsaw Institute for Economic Studies
- Maciej Burny, Head of Regulations, PGE Capital Group

- Stanisław Cios, Energy Policy Unit, Ministry of Foreign Affairs
- Tomasz Dąbrowski, Director, Ministry of Economy, Energy Department (phone)
- Ewa Florkiewicz, EU Economic Department, Ministry of Foreign Affairs
- Andrzej Kassenberg, Head of Institute for Sustainable Development
- Jan Mizak, Ministry of the Environment, Director, Department of Sustainable Development
- Lidia Puka, expert, Polish Institute of International Affairs
- Marcin Ścigan, Expert, Ministry of Economy, Renewable Energy Department
- Dominik Smyrgała, expert, Collegium Civitas
- Agnieszka Sosnowska, Ministry of the Environment, Deputy Director, Department for Air Protection
- Artur Wieczorek, Green Party Secretary

CHAPTER 8

- Maurits Blanson Henkemans, Ministry of Economic Affairs, 11 June 2014
- Ton van Dril, Energy Research Centre of the Netherlands, 10 June 2014
- Willem de Goede, Confederation of Netherlands Industry and Employers, 13 June 2014 (telephone interview)
- Jan Hendriks, Ministry of Economic Affairs, 11 June 2014
- Dennis Holtrop, Ministry of Infrastructure and the Environment, 11 June 2014
- Sander Lensink, Energy Research Centre of the Netherlands, 10 June 2014
- Jos Notenboom, PBL Netherlands Environmental Assessment Agency, 11 June 2014
- Martijn Verdonk, PBL Netherlands Environmental Assessment Agency, 11 June 2014

CHAPTER 9

- Unni Berge, former representative of Zero. Oslo, 26 May 2014
- Agnethe Dahl, Ministry of Climate and the Environment, telephone interview, 5 September 2014

- Paal Frisvold, former representative of Bellona, telephone interview, 2 June 2014
- Mari Hegg Gundersen, Norwegian Water Resources Directorate, Oslo, 5 June 2014
- Knut Kroepelien, former representative of the Ministry of the Environment, at the Norwegian Delegation to the EU, Oslo, 27 May 2014
- Kjetil Lund, former State Secretary, Ministry of Finance, Oslo, 26 August 2014
- Dag Svarstad, Ministry of the Environment, Oslo, 26 May 2014

Index

ACEA (Automobile Manufacturers Association) 63, 81
administrative organization, domestic 4, 20, 22–3, 219, 222, 223
  in Germany 112–13, 116, 221
  in the Netherlands 166–7, 171
  in Norway 192–3
  in Poland 137–9, 141, 220
AEAs (Annual Emissions Allocations) 79, 82, 127, 206, 233
allowances allocation 38, 39, 45, 58, 76, 77, 205–6, 209, 231, 232–3, 237n, 245
  auctioning 12, 38, 67, 72, 78, 82, 85
  in Germany 93, 95, 97–8, 105, 109, 110
  in the Netherlands 150, 151, 152, 168
  in Norway 180–81, 187
  in Poland 122, 123, 124, 125–6, 135, 139, 140
ALTENER programme 33, 34, 36
Ancygier, A. 76, 123, 124, 133, 135, 136, 137, 138, 140
Andersen, M.S. 147
'An Inconvenient Truth' 54

Balkenende government 150, 162, 215
Barroso, J.M.D. 50, 53, 58, 59, 60, 75, 81, 169
BDI (Federation of German Industries) 110, 114
BEE (German Renewable Energy Industry Association) 114
Belchatow power plant 129
Berlusconi, S. 76
biofuels 41–2, 56, 61, 63–4, 67n, 73–4
  Dutch targets 151, 155, 156
  EU 2020 package issues 79–80, 83, 86
  German targets 93, 96, 101, 103, 106
  targets for 41–2, 56, 64
Biofuels Quota Act 101
biomass 52, 56, 135, 148, 184, 217
  co-firing of 121, 123, 125, 128–9, 132, 139, 141, 142, 149, 156, 157, 158, 159–60, 205, 206, 207, 210, 211
Birchfield, V.L. 31
BMWi (Bundesministerium für Wirtschaft und Energie) 91–2, 93, 111, 112
Boasson, E.L. 95, 176, 177, 184
Bøeng, A.C. 79, 183
Bokwa, A. 121
Boratyński, J. et al. 135, 140
Brink, P. ten 81
Buehler, R. 113
Bukowski, M. 134, 135
Bulmer, S. et al. 93
burden-sharing 33, 34, 36, 37, 44, 63, 64, 72, 73–4, 96, 148, 168, 222, 230, 235, 246

carbon leakage 36, 230, 232, 233
  and EU 2020 package 61, 72, 77, 82, 84
  in Germany 105, 114
  in the Netherlands 152, 153
  in Norway 181, 188, 190, 191, 192, 194, 197
  in Poland 123, 126
carbon pricing 17, 38, 209
  and EU 2020 package 54, 61, 74, 78, 85, 86
  and EU 2030 forecasts 232–3
  in Germany 98, 104, 105, 225
  in the Netherlands 163, 168
  in Norway 187, 188, 190

in Poland 123, 124, 130, 135, 139, 141, 142, 216, 224
carbon reduction targets *see* CO_2 emission reduction
Car Emissions Regulation 6–7, 62, 63, 64, 66, 182, 233, 243
in Germany 94, 96, 99–100, 106
negotiations for 80–81, 83, 86, 87, 94, 96
in the Netherlands 155
in Norway 182, 186, 188, 189, 190, 191, 192, 194, 197
CCS (Carbon Capture and Storage) 207, 208, 233
and domestic politics 218, 219, 220, 221
and EU 2020 package 54, 58–9, 61–2, 64, 65, 71, 74, 78, 243
fit and adaptation pressure 213
in Germany 95, 103–4, 105, 108
in the Netherlands 149, 151, 158–9, 161, 163, 167
in Norway 177, 184–6, 188, 190–91, 193, 195
in Poland 124, 129–30, 132, 136–7, 246
transposition of 205
CDM (Clean Development Mechanism) credits 39, 72, 76, 77, 80, 195, 196, 206
CHP (combined heat and power) 41, 94, 101, 135, 149, 164
Christensen, A.R. 62, 63, 80, 81
Christiansen, A.C. 42
Clean and Efficient (Schoon en Zuinig) 150
Climate Agenda for 2030 (Netherlands) 154, 160
Climate Policy Communication 61
CO_2 emission reduction 32, 34, 38, 242
and EU 2020 package 55, 57, 59, 61–3, 78, 80, 81, 84
in Germany 94, 99–100, 103–4, 108, 113
in the Netherlands 153, 154–5, 158–9, 160, 168, 220
in Norway 175, 176, 177, 181, 185, 192, 193, 218
in Poland 120, 124, 127, 129, 131
reduction of 57, 80
trading of 38
co-firing 143n
in the Netherlands 149, 156, 157, 159–60
in Poland 121, 123, 125, 128–9, 139, 207, 210, 211
Communications (European Commission) 32, 36, 40, 74, 83, 168
on 'Building a global carbon market' 58
on CCS policies 61–2
on Climate Policy 61
Energy Communication (2007b) 61
on Energy Efficiency 52
on Energy Union Package 249
on European Energy Security Strategy 229
Winning the Battle against Global Climate Change 51
consensual approach 22, 111, 115, 137, 164–6, 218–19, 222, 249
see also poldermodel
consensus building 13, 107, 162, 247
Copenhagen Summit 72, 75, 76, 82, 132
Costa, O. 5, 17

Davies, C. 78
Day, C. 53, 59
decarbonization *see* CO_2 emission reduction
Delors, Jacques 33, 34
DG Climate Action 83, 84, 85
DG Competition 53, 78, 82
DG Energy 32, 35–6
Energy Roadmap 84
DG Enterprise 53, 58, 60
DG Environment 32, 38, 42, 43, 45, 52–3, 58–9, 60, 61, 62, 64, 66, 71, 244
DG Taxation 33
DG TREN 40–45, 52, 53, 58–9, 60, 64, 66, 71, 244, 252, 253
Green Paper review 54, 57
Di Lucia, L. 4, 19
Dimas, Stavros 50, 53, 55, 75

distribution *see* burden-sharing
domestic politics 4, 20–23, 246, 248
in Germany 107–10
in the Netherlands 162–7, 215, 217, 219, 220–21, 222, 223
in Norway 189–94, 215, 216, 217–18, 219, 220, 221, 222, 223
in Poland 133–9, 215, 216, 217, 220, 222, 223
Doyle, A. 78
Dreger, J. 59, 65
Duff, A. 53
Duffield, J.S. 31, 91–2
Dupont, C. 209
Dür, A. et al. 2, 12
Durban Platform for Enhanced Action 82

ECCP (European Climate Change Programme) 38
ECCP II 51, 52–3
Ecological Tax Reform 93
Economist, The 109
EEA (European Economic Area) Agreement 6, 175, 178, 179–80, 182, 185, 194
EED (Energy Efficiency Directive) 85, 228, 229, 234
Egenhofer, C. 15
employment 3, 12, 60, 91, 109, 114, 136, 139, 142, 152, 164, 176, 216, 225
Energiewende 97, 112, 113
Energy Agreement for Sustainable Growth ('Energy Agreement') (Netherlands) 152–3, 154, 157, 159, 164, 171, 219
Energy Communication (2007b) 61
Energy Efficiency Communication 52
Energy Performance of Buildings Directive 41, 127
Energy Product Tax Directive 41
Energy Union 229, 249
Environmental Management Act (Netherlands) 154
EREC (European Renewable Energy Council) 63
ESA (EFTA Surveillance Authority) 176, 180, 181, 185, 186, 219
ESD (Effort-Sharing Decision) 206, 208
and EU 2020 package proposal 71, 72–3, 78–9
fit and adaptation pressure 213
in Germany 95, 98–100, 106
and low-carbon solutions 232–3
in the Netherlands 153–5
in Norway 181–3, 186, 188, 190–91, 192
in Poland 126–7, 132
transposition of 206
ETS (emissions-trading system) 38–40, 44–5, 206, 208, 242
and domestic politics 217, 219
and EU 2020 package proposal 71–85, 94–5
and EU 2030 framework 230–31
fit and adaptation pressure 212
in Germany 97–8, 105
long-term views 209
and low-carbon solutions 232
in the Netherlands 149, 152–3, 160
in Norway 180–81, 186, 188, 190, 191, 192
in Poland 123, 125–6, 131
transposition of 204
EU 2020 package proposal 86–7
directives for 71–5
international committment 81–5
negotiation for 75–81, 94–6, 243–5
EU 2030 targets 2, 4, 8, 23–4, 25, 61, 84, 204, 227–31, 250
and Germany 97, 99, 113, 114, 116
and low-carbon solutions 233–4
and the Netherlands 154–5, 160, 164, 168–9, 225, 226, 227
and Norway 194–6, 197, 225, 226
and Poland 14, 127, 139–40, 224–5, 226, 227–8, 229–30
Eurelectric 38
Eurobarometer 54
European Alliance of Energy Intensive Industries 83
European Coal and Steel Community Treaty (1951) 31

European Commission
on carbon leakage 72
decision-making role 14–15
energy directives 41–2
on ETS 38–9, 58, 83–4
on EU 2030 targets 228–9
policy packages 32–4
Reasoned Opinion to Poland 125, 128
on state-aid rules for ETS revision 78
European Council 40–41, 51–2, 175
on burden-sharing 63
decision-making role 15–16
and DG Energy Agenda 35–6
on EU 2020 package 73, 75, 77, 78, 80–81, 83–4, 86, 122, 133, 150
on EU 2030 framework 142, 168, 229–30, 231, 233, 234
and EU 2050 targets 120
and Green Paper review 55
and tax proposal 33
European Energy Security Strategy Communication 229
European Environment Agency 91, 99, 126, 127, 148, 152, 153, 154, 155, 157, 205, 206, 233
European Parliament
on CO_2 emissions 124
decision-making role 16, 38, 39
on EU 2020 package 50, 54, 55–8, 62–3, 75, 77, 78–80
and EU 2030 framework 228, 232
European Strategic Energy Technology Plan 61
Europia 63

Fairbrass, J. 14
Federation of German Industries (BDI) 110, 114
feed-in tariff system 234
in Germany 93, 101–2, 109–10, 112, 116, 206–7, 217
in the Netherlands 148, 151, 155
in Poland 128, 135, 136
fit and adaptation pressure 4, 19–20, 26, 212–14, 235, 246, 247, 248
in Germany 105–6, 115
in the Netherlands 160–61, 170
in Norway 187–9, 190
in Poland 130–33, 141
FQD (Fuel Quality Directive) 7, 62, 64, 66, 209, 222, 233, 243, 245
EU 2020 package negotiations 80–83, 87
in Germany 96, 99, 101, 106
in the Netherlands 168
in Norway 182, 186, 188, 190, 191, 192, 193, 194, 197
France 33, 56–7, 60, 64
EU 2020 package proposals 76, 83

Gabriel, S. 94, 112
gas supply 6, 56, 61, 92, 176, 195, 229
Dutch 148, 161, 163
Russian 53, 66, 121
geothermal energy 157, 158
Germany 6, 115–16, 163
consequences of implementation 113–15
domestic politics 105–6, 107–12
energy policy 56–7, 92–3
and ETS reform 39, 77–8, 86
EU 2020 package proposal 94–105
fit and adaptation pressure 105–6
taxation 93
transport policies 80–81
Germany, implementation analysis 205, 206–7, 208
domestic politics 215–16, 217, 218, 220–21, 222, 223
and EU 2030 framework 225
fit and adaptation pressure 212, 213
long-term views 209, 210
GHG emission reduction 1, 5, 8n, 18, 32, 34, 37, 38, 43–4, 209, 246
EU 2020 targets 51, 55, 60–61, 63, 72, 77, 79–80, 81–2, 83–4, 205, 208
EU 2030 framework 227–8, 230, 232
in Germany 91, 94–9, 104, 106, 107, 113–14, 115, 211–12
in the Netherlands 147, 148, 150, 152–5, 162, 168–70, 215
in Norway 180, 188
in Poland 120, 122–4, 126–7, 140, 142

goodness-of-fit *see* fit and adaptation pressure
Gore, Al 54
green certificates scheme
  in the Netherlands 148
  Norwegian/Swedish 177, 183–4, 186, 188, 190, 192, 193, 206, 207, 210
  in Poland 121, 128, 141, 206
Green European Foundation 158
'Green Growth Group' 6, 168, 229
Greenhouse Gas Emission Trading Act 97
'green jobs' 3, 12, 114, 139, 142, 216–17, 224, 225, 244
Green Paper 38, 39, 179
  on 2030 framework 168, 227, 229
  on Strategic Energy Review 54–8
Green Party (Germany) 92, 111
Guardian, The 133, 169

Hampton Court Summit 52, 66
Hancher, L. 32
Hanf, K. 4
Hasselknippe, H. 42
heating sector 42, 54, 79, 82
  German targets 94, 101, 103, 104, 115, 207
  negotiations for 94, 100
  in Norway 175, 179
Hedegaard, C. 83, 228
Heine, D. et al. 93
Hekkenberg, M. 157, 158
High-Level Group 53, 55, 66
Hix, S. 13, 14
Hooghe, L. 3, 13
Hovi, J. 11
Howes, T. 79, 151
Hudson, V.M. 21
hydropower 6, 176, 184, 186, 210

IEA (International Energy Agency) 129, 156–7, 159–60, 161, 165
Impact Assessments System 53
internationalization 4–5, 16–17, 19, 25, 32–3, 35, 37, 151, 153, 169, 245, 250
  on 2020 targets 53–4, 60, 66, 74, 75, 76, 81–7
IPCC (Intergovernmental Panel on Climate Change) 1, 32, 51
issue-linkage mechanisms 2–3, 5, 11–15, 26, 65, 66, 85, 214, 241, 244–5, 247–8
  in Poland 139
Italy 42, 76, 83, 138, 228
ITRE Committee 57, 79

Jendroska, J. 129, 137, 138
Jordan, A. 14
Jørgensen, K.E. 5, 17

Kaczynski, L. 122, 133
Katzenstein, P. 92
Kedzierzyn SA plant 129
Kelly, C.R. 34, 37
Knill, C. 17, 19
Koelemeijer, R. et al. 168
Kopacz, E. 142
Kronsell, A. 4, 19
Kyoto Protocol 37, 39, 45, 51, 54, 66, 87, 148, 242, 244, 245
  conflicts 42–4

Lacasta, N. 79
Langhelle, O. 177
Lauber, V. 36, 93
'Law of the Least Ambitious Programme' 11
Lenschow, A. 17, 19
Liberal Intergovernmentalism (LI) 3, 13–15, 23, 26, 45, 66, 86, 235, 244, 245, 247, 248
Liefferink, D. 147
Lijphart, A. 164
'Linking Directive' (Directive 2004/101/EC) 39, 122
Lisbon European Council 40
low-carbon transition 4, 12, 24, 27n, 51, 54–5, 204, 209–10, 211, 231–3
  and EU 2020 targets 71–4, 76, 77, 83, 84
  in France 60, 64
  in Germany 92, 105, 107, 112, 113, 116

in the Netherlands 147, 148, 152, 163, 168, 169, 170, 217
in Norway 195
in Poland 120, 126, 129, 130, 135, 139, 140, 225
Loyola de Palacio, Ignacia 42–4
LTA3 (Long-Term Agreement) 154
LULUCF (land use and land-use change and forestry) 233
Lyons, P.K. 32

majority government 21, 38, 39, 108, 133, 134, 162, 215
Malnes, R. 14
Maloney, W.A. et al. 22
Market Incentive Programme 101, 103
Market Stability Reserve 113, 114, 232
Marks, G. 3, 13, 14
McKibben, H. 12, 15, 16
Merkel, A. 64, 81, 97, 112
government 97, 99, 101, 102, 103, 107, 108, 109
Mez, L. 93
Ministry of Infrastructure and the Environment (Netherlands) 152, 162, 167, 171
Ministry of Petroleum and Energy (Norway) 176, 183, 185, 192–3
Moravcsik, A. 3, 13
Motor Vehicle Tax Act (Germany) 99
MSR agreement 232
Multi-level Governance (MLG) 3, 14, 16, 19, 26, 45, 66, 244, 245, 248
Müngersdorff, M. 79–80

NAP (National Allocation Plan) 39, 54, 58, 72, 105, 122
National Development Plan for Electric Mobility 99
National Environmental Policy Plan (Netherlands) 148
NCPIP (National Climate Policy Implementation Plan) 148
negotiations 1–4, 6, 11–15
of EU 2020 package proposal 75–81, 243–5
in Germany 94–6
in the Netherlands 150–52
in Norway 178–9, 184
in Poland 122–5
on promotion of renewables 40–42
on tax conditionality 33–4
theory 2–3, 11, 245, 247
see also issue-linkage mechanisms
NER 300, 82, 86, 95, 104, 110, 114, 130, 159, 207, 228, 231
NER 400, 114, 231, 233
Netherlands 147, 159–60, 168–71
CCS Directive 149, 151, 158–9, 161
domestic politics 162–7
energy import 6
ESD policies 153–5
ETS policies 149, 152–3, 160
negotiations in 150–52
policies up to 2007, 147–9
RED policies 151, 155–8, 160–61
Netherlands, implementation analysis 204, 206–7, 208
domestic politics 215, 216, 217–18, 219, 220–21, 222, 223
EU 2030 framework 225, 226, 227
fit and adaptation pressure 212, 213, 214
long-term views 209, 210, 211, 212
Nilsson, M. 41, 65
NIMs (national implementation measures) 208
in Germany 98, 105, 205
in the Netherlands 152
in Norway 180, 181
in Poland 126
Norway 6, 7, 175, 186–7, 196–7
CCS Directive 177, 184–6
consequences of policy implementation 194–6
domestic politics 189–94
ESD policies 181–3
ETS policies in 180–81
fit and adaptation pressure 187–9
negotiations in 178–9
policies up to 2007, 176–7
RED policies 183–4
Norway, implementation analysis 206, 207, 208
domestic politics 215, 216, 217–18, 219, 220, 221, 222, 223

EU 2030 framework 225, 226
fit and adaptation pressure 212, 213
long-term views 209, 210
Norwegian Oil Industry Association 176, 192
Notenboom, J. et al. 159
NREAP (National Renewable Energy Action Plan)
in Germany 101
in the Netherlands 156, 157
in Norway 184
nuclear power 63, 84
in France 33, 60
phase-out in Germany 92, 97, 107, 109, 114
in Poland 121, 133
Nugent, N. 16

Oberthür, S. 34, 37, 209
Oettinger, G. 102, 228
Olsen, J.P. 17

Padgett, S. 32
Paris Summit 114, 140, 225, 228, 229, 230, 250
PBL (Planbureau voor de Leefomgeving) 151, 168
Peeters, M. 154
Piebalgs, A. 50, 52, 53, 59, 75
Piria, R. et al. 225
PKEE 124
Poland 120
and CCS Directive 124, 129–30, 132, 136–7, 246
domestic politics 133–9
electricity production 74
energy policy 56–7
ESD policies 126–7
ETS policis 76–7, 83, 84, 123, 125–6
fit and adaptation pressure 131–3, 141, 142
implementation policies 6, 139–40
NAP failing 122
package negotiations 122–5
RED policies 121, 124, 127–9
Poland, implementation analysis 204–5, 206, 207, 208, 235
domestic politics 215, 216, 217, 220, 222, 223
EU 2030 framework 224–5, 226, 227–8, 229–30
fit and adaptation pressure 212, 213, 214
long-term views 209, 210, 211
poldermodel 165, 222
policy phases approach 11–12, 15, 26
Polish Energy Policy 134, 136, 229
political involvement
in Germany 92, 93, 97, 103–4, 107, 111–12
in the Netherlands 162, 164, 168
in Poland 133–4
pollution, air 12, 24, 92, 139, 142
Prodi, R. 43

qualified majority voting 14, 27n, 37, 38, 39, 45, 76, 242

Rathmann, M. et al. 149, 157, 158
Reasoned Opinion 125, 128
RED (Renewable Energy Directive) 65, 71, 206–7, 208
and domestic politics 215, 217–18, 219, 220
and EU 2020 policy package 243–4
fit and adaptation pressure 212, 213
in Germany 100–103, 105–6
long-term views 210
in the Netherlands 151, 155–8, 160–61
in Norway 183–4, 186, 188, 190, 192, 193, 194
in Poland 124, 127–9, 131–3, 136
targets 73–4, 79–80
transposition of 204–5
Regulation on emission performance standards for new passenger cars (EU 443/2009) 127, 155
Regulation on State aid to the coal industry (EC 1407/2002) 41
renewable energy 33, 34, 36–7
Energy Roadmap 2050, 84
and ETS reform 62
in Germany 94, 96, 97, 100–103, 108–9

in the Netherlands 148
promotion of 40–42, 43, 52, 54, 59, 64
taxation 93
see also RED
Renewable Energy Heating Act 101, 103
RES (Renewable Energy Standard) proposal 151, 158
Richardson, J.J. 3, 14
Ringius, L. 33, 36
Rio Conference (1992) 31, 33, 34, 44, 242
Ripa di Meana, Carlo 34
ROAD project 159
Rutte government 155, 162, 169

Sarkozy, N. 77, 81, 123
SAVE programme 33, 34
Sbragia, A.M. 15
Schmidt, S. 113
SDE-plus scheme *(Stimulering Duurzame Energieproductie)* 155–6, 158, 160, 162
Sebenius, J.K. 2, 11, 12
security-of-supply 31, 35, 40, 53–7, 227
in Germany 92
in the Netherlands 161
in Norway 176, 189, 195
in Poland 132–3
SER (Social and Economic Council) 152, 165
Shell 152, 163, 168, 217
Barendrecht project 158–9, 167
side-payments 3, 12, 24, 245, 247
and EU 2020 package proposal 74, 86
in Germany 109, 116
in Poland 139
Single European Act (1986) 32
Skidmore, D. 21
Skodvin, T. 11, 38
'Small Tri-Pack' Act 128
solar PV 101–2, 156
solidarity funds 76–77, 123–4, 231
sovereignty 14, 42, 55, 57, 140, 229
Soviet Union, supply from 53, 121
Spain 36, 42, 43, 138, 237n
state-aid guidelines 216, 222, 234
and EU 2020 package proposal 75, 78, 82
in Germany 95, 102
in Norway 176
in Poland 126
Statoil 182, 185, 186
Stern Review 60
Strategic Energy Review Green Paper 54, 55
Strategy on Climate Policy En Route to 2020, 153
Sustainability Agenda *(Agenda Duurzaamheid)* 154
Sweden 79, 176–7, 183–4, 187, 188, 207, 220
Szulecki, K. 138

taxation 33–5, 76, 143n, 198n, 242
Energy Tax Directive 41, 42
in Germany 93, 98, 99, 100, 101, 103, 109
in the Netherlands 148, 157, 158
in Norway 177, 181, 182, 188, 192, 218
Tjernshaugen, A. 177
Tollison, R.D. 12
transport sector 6–7, 41, 62, 63
EU 2020 package 80–81, 82–3, 242
in Germany 94, 96, 99–100, 101, 103, 106
long-term views 209–10
in the Netherlands 151, 154–5
in Norway 182, 188, 190–91, 192
in Poland 126–7
see also Car Emissions Regulation
transposition 18, 21, 98–9, 204–5, 206, 208, 211, 245–6
CCS 103, 104, 108, 115, 129–30, 132, 138, 158
ETS 97–8, 125, 152, 180, 186
of the 'Linking Directive' 122
RED 100–101, 104, 128, 132, 155, 183, 193
Treib, O. 4, 17, 19
Tsebelis, G. 12, 21
Tusk, D. 123, 133–4, 142, 229

UNCED (UN Conference on Environment and Development) 33, 34
Underdal, A. 3, 4, 11, 14, 21
UNFCCC (UN Framework Convention on Climate Change) 17, 34, 36, 54, 122
United Kingdom
  and climate-change goals 33, 59–60
  energy policy 56
  Hampton Court Summit 52, 66
  security-of-supply policy 31
  and tax proposal 34
United States, and Kyoto Protocol negotiations 37, 39

Vattenfall AG 104
VDMA (German Engineering Association) 114
Verdonk, M. 157, 158
Verheugen, Günther 51, 53, 59
veto players 13, 20, 21, 111, 112, 171, 231
veto power 16, 134
  in Germany 107–9, 113
  Polish veto 76, 84, 86, 120, 140, 225, 229
Visegrad Group (V-4) 76, 123, 140
Vitols, S. 92
VNO-NCW (Confederation of Netherlands Industry and Employers) 163, 168
VROM (Ministry of Housing, Spatial Planning and Environment) 149
VVD (Volkspartij voor Vrijheid en Democratie) 168

Wallström, M. 43–4
Westphal, K. 91–2
Wettestad, J. 3, 34, 38, 39, 40, 50, 54, 58, 76, 77, 95, 122, 123, 149
Wicken, O. 176
Willett, T.D. 12
Wilson, J.Q. 21
windpower 143n
  in the Netherlands 157, 158, 164, 167, 220
  in Norway 176, 183, 184
  in Poland 128, 129, 135, 138
Winning the Battle against Global Climate Change 51

Ydersbond, I.M. 122
Young, O. 14

Ziegler, N. 93